W0018251

Advanced Courses in Mathematics
CRM Barcelona

Centre de Recerca Matemàtica

Managing Editor:
Manuel Castellet

Vesselin Drensky
Edward Formanek

Polynomial Identity Rings

Birkhäuser Verlag
Basel · Boston · Berlin

Authors' addresses:

Vesselin Drensky
Institute of Mathematics and Informatics
Bulgarian Academy of Sciences
1113 Sofia
Bulgaria
drensky@math.bas.bg

Edward Formanek
Department of Mathematics
Pennsylvania State University
University Park, PA 16802
USA
formanek@math.psu.edu

2000 Mathematical Subject Classification 16R, 15A24, 15A72

A CIP catalogue record for this book is available from the
Library of Congress, Washington D.C., USA

Bibliografische Information Der Deutschen Bibliothek
Die Deutsche Bibliothek verzeichnet diese Publikation in der Deutschen Nationalbibliografie; detaillierte
bibliografische Daten sind im Internet über <http://dnb.ddb.de> abrufbar.

ISBN 3-7643-7126-9 Birkhäuser Verlag, Basel – Boston – Berlin

This work is subject to copyright. All rights are reserved, whether the whole or part of the material is concer-
ned, specifically the rights of translation, reprinting, re-use of illustrations, recitation, broadcasting,
reproduction on microfilms or in other ways, and storage in data banks. For any kind of use permission of the
copyright owner must be obtained.

© 2004 Birkhäuser Verlag, P.O. Box 133, CH-4010 Basel, Switzerland
Part of Springer Science+Business Media
Cover design: Micha Lotrovsky, 4106 Therwil, Switzerland
Printed on acid-free paper produced from chlorine-free pulp. TCF∞
Printed in Germany
ISBN 3-7643-7126-9

9 8 7 6 5 4 3 2 1   www.birkhauser.ch

Contents

B   Polynomial Identity Rings
Edward Formanek   **131**

Foreword

Algebras satisfying polynomial identities, called PI-algebras, have been one of the major research fields in the last 20 years, since they constitute a reasonably big class containing the finite dimensional algebras and the commutative algebras and enjoy many of their main properties.

These notes correspond to the lectures delivered by the authors in the Advanced Course on Polynomial Identity Rings, held at the Centre de Recerca Matemàtica in Bellaterra (Barcelona), July 2003. One strand of lectures concentrated on the combinatorial side of polynomial identity rings (PI-rings), while a second explained the structural side of PI-rings. Thus these notes, which are revised and smoothed version of the lecture notes delivered to the participants of the Advanced Course, are organised in two parts corresponding to these two lecture series.

The first part, due to V. Drensky, is devoted to introducing the reader to the combinatorial aspects of PI-theory. The purpose of this chapters is to present several results which form the foundation of the combinatorial theory of PI-algebras, including a survey on some recent results on related topics.

The second part, due to E. Formanek, is an introduction to the structural aspects of polynomial identity rings, including the Amitsur–Levitzki theorem, central polynomials, the theorems of Kaplansky, Posner and Artin, and the ring of generic matrices.

Besides our indebtedness to the Centre de Recerca Matemàtica, thanks are due to Ferran Cedó, the course co-ordinator, for making it possible, and to the CRM secretaries, Consol Roca, Maria Julià and Neus Portet, for their assistance. Special thanks go to all the participants of the course for their interest in the event and for their very helpful contributions during the afternoon sessions.

E. Formanek
V. Drensky

Part A

Combinatorial Aspects in PI-Rings

Vesselin Drensky

Introduction

The class of PI-algebras, i.e., algebras satisfying polynomial identities, is a reasonably big class containing the finite dimensional and the commutative algebras and enjoying many of their properties. Traditionally the theory of PI-algebras has two aspects, structural and combinatorial, with considerable overlap between them.

The purpose of these lecture notes is to present several results which form the foundation of the combinatorial theory of PI-agebras: The Amitsur–Levitzki theorem; the construction of central polynomials for matrices; the polynomial identities of matrices and their relation with invariant theory; the Nagata–Higman theorem on the nilpotency of nil algebras of bounded index; the Shirshov theorem for finitely generated PI-algebras; the Regev theorem for the tensor product of PI-algebras.

In many places in the text a survey on some recent results on related topics is given, such as the negative solution of the Specht problem in positive characteristic, applications of computers to the polynomial identities of matrices, central polynomials of low degree for matrices of any size, growth of PI-algebras and their codimension sequences. Also, a couple of results which usually are not included in courses on PI-algebras are presented, such as the lower bound of Kuzmin for the class of nilpotency in the Nagata–Higman theorem. I hope that the included open problems, comments and references to the current situation in the discussed field will be useful for the orientation of the reader.

I tried to make the exposition accessible for young mathematicians with standard background on linear algebra and elements of ring and group theory. Nevertheless I believe that professional mathematicians working on PI-theory also will find the text useful.

There are several books which can serve for further reading on some of the topics included in these lecture notes. Among them are the general books on PI-algebras by Procesi [Pr2], Jacobson [J] and Rowen [Rw1]. The books by Formanek [F6], Kemer [Ke5] and Razmyslov [Ra5] are devoted to more specific topics. Recently, I wrote the book [Dr10] which deals with combinatorial theory of free algebras and PI-algebras. I tried to minimize its intersection with the present text.

Acknowledgements

I am very grateful to Ed Formanek for the useful discussions and suggestions about the topics of the course. I am also very thankful to many colleagues and friends. I want especially to mention my advisers Georgy Genov from the University of Sofia and Yuri Bahturin from the Moscow State University, as well as S.A. Amitsur, L.A. Bokut, O.M. Di Vincenzo, M. Domokos, M.B. Gavrilov, A. Giambruno, A.R. Kemer, P. Koshlukov, V.N. Latyshev, A.P. Popov, Yu.P. Razmyslov, A. Regev, I.P. Shestakov, A.L. Shmelkin, I.B. Volichenko, and E.I. Zelmanov.

Finally, I am deeply indebted to my family and especially to my wife for the family atmosphere, the support and the understanding as long as I was working on this text.

This project was also partially supported by Grant MM-1106/2001 of the Bulgarian Foundation for Scientific Research.

Chapter 1

Basic Properties of PI-algebras

Throughout this part devoted to the combinatorial aspects in PI-rings we fix a field K. All algebras which we consider will be associative and over K. Most of the algebras will be unitary. We shall mention it explicitly, when we consider nonunitary algebras. All vector spaces, tensor products, etc. will be also over K. Although some of the results hold over fields of any characteristic, if not explicitly stated, we assume that the characteristic of K is equal to 0.

1.1 PI-algebras – Definitions and Examples

For a commutative algebra R,

$$rs - sr = 0 \quad \text{for any two elements} \ r, s \in R.$$

Similarly, if the (nonunitary) algebra R is nilpotent of class n, then

$$r_1 \cdots r_n = 0 \quad \text{for all} \ r_1, \ldots, r_n \in R.$$

For a finite dimensional algebra R of dimension $< n$, let us consider the expression

$$s_n(r_1, \ldots, r_n) = \sum_{\sigma \in S_n} (\text{sign} \ \sigma) r_{\sigma(1)} \cdots r_{\sigma(n)}, \quad r_1, \ldots, r_n \in R,$$

where the summation runs on all permutations in the symmetric group S_n of degree n acting on the first n integers $1, \ldots, n$. The expression $s_n(r_1, \ldots, r_n)$ is skew-symmetric and behaves as a determinant. Since $\dim R < n$, the elements $r_1, \ldots, r_n$ are linearly dependent and $s_n(r_1, \ldots, r_n) = 0$.

We shall introduce the formal generalization of the above examples.

Definition 1.1.1. For every set X the algebra $K\langle X \rangle$ with basis consisting of all words

$$x_{i_1} \cdots x_{i_n}, \quad x_{i_j} \in X, \quad n = 0, 1, 2, \ldots,$$

and multiplication defined by

$$(x_{i_1} \cdots x_{i_m})(x_{j_1} \cdots x_{j_n}) = x_{i_1} \cdots x_{i_m} x_{j_1} \cdots x_{j_n}, \quad x_{i_k}, x_{j_l} \in X,$$

is the *free associative algebra* freely generated by the set X. We shall call the elements of $K\langle X \rangle$ polynomials in the noncommuting variables X.

We shall also consider the *free nonunitary algebra* $K^+\langle X \rangle$ which consists of all polynomials from $K\langle X \rangle$ without constant terms.

The following lemma gives the *universal property* of the free algebras. Its proof is an easy exercise.

Lemma 1.1.2. *Let R be any algebra and let $\varphi_0 : X \to R$ be any mapping. There exists a unique homomorphism $\varphi : K\langle X \rangle \to R$ which extends φ_0, i.e., $\varphi|_X = \varphi_0$.*

From now on we fix a countable infinite set $X = \{x_1, x_2, \ldots\}$. Sometimes we shall also use other symbols, (e.g., y, z, y_i, z_j, etc.) for the elements of X.

Definition 1.1.3. (i) Let $f = f(x_1, \ldots, x_m) \in K\langle X \rangle$ and let R be an algebra. We say that $f = 0$ is a *polynomial identity for R* if

$$f(r_1, \ldots, r_m) = 0 \quad \text{for all} \quad r_1, \ldots, r_m \in R.$$

Sometimes we shall also say that f itself is a polynomial identity for R.

(ii) If the algebra R satisfies a nontrivial polynomial identity $f = 0$ (i.e., f is a nonzero element of $K\langle X \rangle$), we call R a *PI-algebra* ("PI" = "Polynomial Identity").

Remark 1.1.4. Clearly, $f \in K\langle X \rangle$ is a polynomial identity for R if and only if f is in the kernel of all homomorphisms $K\langle X \rangle \to R$.

The motivating examples in the beginning of the section can be restated as follows.

Examples 1.1.5. (i) The algebra R is commutative if and only if it satisfies the polynomial identity

$$[x_1, x_2] = x_1 x_2 - x_2 x_1 = 0.$$

The expression $[x_1, x_2] = x_1 x_2 - x_2 x_1$ is called the *commutator* of x_1 and x_2. Inductively one defines *longer commutators* by

$$[x_1, \ldots, x_{n-1}, x_n] = [[x_1, \ldots, x_{n-1}], x_n], \quad n = 3, 4, \ldots.$$

(ii) The algebra R is nilpotent of class of nilpotency $\leq n$ if and only if it satisfies the polynomial identity $x_1 \cdots x_n = 0$. Of course, R is nonunitary and we have to consider $x_1 \cdots x_n$ as an element of the free nonunitary algebra $K^+\langle X \rangle$.

(iii) If R is a finite dimensional algebra and dim $R < n$, then R satisfies the *standard identity of degree n*

$$s_n(x_1, \ldots, x_n) = \sum_{\sigma \in S_n} (\text{sign } \sigma) x_{\sigma(1)} \cdots x_{\sigma(n)} = 0.$$

The algebra R also satisfies the *Capelli identity*

$$d_n(x_1, \ldots, x_n; y_1, \ldots, y_{n-1}) = \sum_{\sigma \in S_n} (\text{sign } \sigma) x_{\sigma(1)} y_1 x_{\sigma(2)} y_2 \cdots y_{n-1} x_{\sigma(n)} = 0.$$

(iv) Since the $n \times n$ matrix algebra $M_n(K)$ is of dimension n^2, it satisfies the standard identity of degree $n^2 + 1$ and the Capelli identity in $n^2 + 1$ skew-symmetric variables. In Chapter 3 we shall prove the Amitsur–Levitzki theorem stating that $M_n(K)$ satisfies the standard identity of degree $2n$. On the other hand, $d_{n^2+1}(x_1, \ldots, x_{n^2+1}; y_1, \ldots, y_{n^2}) = 0$ is the Capelli identity of minimal degree which holds for $M_n(K)$.

(v) By the Cayley–Hamilton theorem every 2×2 matrix r satisfies the equation

$$r^2 - \operatorname{tr}(r)r + \det(r)e = 0,$$

where e is the identity matrix. If $\operatorname{tr}(r) = 0$, then $r^2 = -\det(r)e$ is a scalar matrix and $[r^2, s] = 0$ for all $s \in M_2(K)$. Since $\operatorname{tr}(r_1 r_2) = \operatorname{tr}(r_2 r_1)$, the easiest way to produce a traceless matrix is to take $r = [r_1, r_2]$. Hence $[[r_1, r_2]^2, s] = 0$ and $M_2(K)$ satisfies the *Hall identity*

$$[[x_1, x_2]^2, x_3] = 0.$$

(This identity appeared already in a paper of Wagner [Wa] in 1937 and has its analogues for matrices of any size.)

Let V be a vector space with basis $\{e_1, e_2, \ldots\}$. The *Grassmann* (or *exterior*) *algebra* $E = E(V)$ is generated by the basis of V and has defining relations

$$e_i e_j + e_j e_i = 0 \quad \text{for all} \ \ i, j = 1, 2, \ldots.$$

In other words, E is isomorphic to the factor algebra

$$K\langle X \rangle / (x_i x_j + x_j x_i \mid i, j = 1, 2, \ldots),$$

under the canonical homomorphism sending x_i to e_i, $i = 1, 2, \ldots$. The defining relations of E allow to rearrange the products of the generators, and

$$e_{\sigma(i_1)} \cdots e_{\sigma(i_n)} = (\operatorname{sign} \sigma) e_{i_1} \cdots e_{i_n}$$

for any permutation σ of the simbols $i_1, \ldots, i_n$. Since $e_i e_i + e_i e_i = 0$ and char $K = 0$, we have also $e_i^2 = 0$. In this way, as a vector space, E is spanned by the products $e_{i_1} \cdots e_{i_n}$, where $i_1 < \cdots < i_n$ and $n = 0, 1, 2, \ldots$. It is well known that these monomials form a basis of E.

Example 1.1.6. The Grassmann algebra E satisfies the polynomial identity

$$[x_1, x_2, x_3] = [[x_1, x_2], x_3] = 0.$$

For the proof, $[x_1, x_2, x_3]$ is linear in each variable x_1, x_2, x_3, and it is sufficient to see that $[r_1, r_2, r_3] = 0$ for the basis elements of E. Direct verification shows that

$$[r_1, r_2] = [e_{i_1} \cdots e_{i_m}, e_{j_1} \cdots e_{j_n}] = (1 - (-1)^{mn}) e_{i_1} \cdots e_{i_m} e_{j_1} \cdots e_{j_n}.$$

Hence $[r_1, r_2] \neq 0$ implies that both m and n are odd integers. Then $[r_1, r_2]$ is of even length and $[[r_1, r_2], r_3] = 0$ for any $r_3 = e_{k_1} \cdots e_{k_p}$.

Example 1.1.7. The algebra $NT_n(K)$ of all $n \times n$ upper triangular matrices with zero diagonal is nilpotent of class n and satisfies the polynomial identity $x_1 \cdots x_n = 0$. Let $UT_n(K)$ be the algebra of all $n \times n$ upper triangular matrices. Then for any $r_1, r_2 \in UT_n(K)$ the commutator $[r_1, r_2]$ belongs to $NT_n(K)$. Hence $UT_n(K)$ satisfies the polynomial identity

$$[x_1, x_2] \cdots [x_{2n-1}, x_{2n}] = 0.$$

1.2   T-ideals and Varieties of Algebras

Definition 1.2.1. Let R be an algebra. The set $T(R)$ of all polynomial identities of R is a two-sided ideal of the free algebra $K\langle X \rangle$ and is called the *T-ideal* of R.

If $f(x_1, \ldots, x_m)$ is a polynomial identity for R, then for any $w_1, \ldots, w_m \in K\langle X \rangle$ the polynomial $f(w_1, \ldots, w_m)$ also vanishes on R. Hence the T-ideals are invariant under substitutions of elements from $K\langle X \rangle$. Since every endomorphism φ of $K\langle X \rangle$ is defined by the image of X, i.e., by $\varphi(x_i) = w_i \in K\langle X \rangle$, the T-ideals are invariant under the endomorphisms of $K\langle X \rangle$. (The ideals U of an algebra R which are invariant under the endomorphisms of R are called *fully invariant*.)

It is easy to see that any fully invariant ideal U of $K\langle X \rangle$ is a T-ideal. For this purpose, let $R = K\langle X \rangle / U$ be the factor algebra of $K\langle X \rangle$ modulo the ideal U. We denote the generators of R by $\bar{x}_i = x_i + U$, $i = 1, 2, \ldots$, and shall show that $U = T(R)$. First, let $f(x_1, \ldots, x_m) \in U$ and let $\bar{w}_1 = w_1 + U, \ldots, \bar{w}_m = w_m + U$ be any elements of R. Since U is fully invariant, the element $f(w_1, \ldots, w_m)$ also belongs to U. This means that $f(\bar{w}_1, \ldots, \bar{w}_m) = f(w_1, \ldots, w_m) + U$ coincides with $\bar{0} = U$ and $f(x_1, \ldots, x_m)$ vanishes under any substitution with elements of R. Hence $U \subseteq T(R)$. On the other hand, if $f(x_1, \ldots, x_m) \in K\langle X \rangle$ and $f \notin U$, then $f(\bar{x}_1, \ldots, \bar{x}_m) \neq 0$ in R and hence $f = 0$ is not a polynomial identity for R. Therefore $U = T(R)$.

Definition 1.2.2. The polynomial identity $g(x_1, \ldots, x_m) = 0$ is called a *consequence of the polynomial identities* $f_i(x_1, \ldots, x_{m_i}) = 0$, $i \in I$, if any algebra satisfying the identities $f_i(x_1, \ldots, x_{m_i}) = 0$ satisfies also $g(x_1, \ldots, x_m) = 0$. We denote by

$$\left(f_i(x_1, \ldots, x_{m_i}) \mid i \in I \right)^T$$

the smallest T-ideal U containing all $f_i(x_1, \ldots, x_{m_i})$, $i \in I$. This T-ideal coincides with the set of all consequences of the identities $f_i = 0$, $i \in I$, and its elements have the form

$$\sum u_{iw} f_i(w_1, \ldots, w_{m_i}) v_{iw}, \quad w_1, \ldots, w_{m_i}, u_{iw}, v_{iw} \in K\langle X \rangle.$$

The generating set $\{ f_i(x_1, \ldots, x_{m_i}) \mid i \in I \}$ is called a *basis* of the T-ideal U, even if it is not a minimal generating set. (Any generating set of the T-ideal $T(R)$ is called a *basis of the polynomial identities* of the algebra R.) Two sets of identities are called *equivalent* if they generate the same T-ideal.

Remark 1.2.3. When we work with not necessarily unitary algebras, we consider the polynomial identities and their consequences in $K^+\langle X \rangle$. The most essential difference to the unitary case is the following. If $f(x_1, \ldots, x_m) \in K^+\langle X \rangle$ is a polynomial identity for an algebra R, then, obtaining consequences, in the nonunitary case we are not allowed to replace the variables x_i with polynomials with constant terms. For example, if $x_1[x_2, x_3] = 0$ is a polynomial identity and we work with unitary algebras, replacing x_1 with 1, we obtain the consequence $[x_2, x_3] = 0$. In the nonunitary case the identity $x_1[x_2, x_3] = 0$ has no consequences of second degree. More generally, in the nonunitary case the elements of the T-ideal with generating set $\{f_i(x_1, \ldots, x_{m_i}) \mid i \in I\}$ have the form

$$\sum u_{iw} f_i(w_1, \ldots, w_{m_i}) v_{iw}$$

with $w_1, \ldots, w_{m_i} \in K^+\langle X \rangle$ and $u_{iw}, v_{iw} \in K + K^+\langle X \rangle$.

The notion of polynomial identity is a special case of the notion of identical relation in abstract algebraic systems, and one can introduce varieties of algebras and related objects.

Definition 1.2.4. (i) Let $\{f_i(x_1, \ldots, x_{m_i}) \mid i \in I\} \subset K\langle X \rangle$. The class $\mathfrak{V}$ of all associative algebras satisfying all polynomial identities $f_i = 0$, $i \in I$, is called the *variety of (associative) algebras defined* (or *determined*) *by the system of polynomial identities* $\{f_i = 0 \mid i \in I\}$. We denote by $T(\mathfrak{V})$ the T-ideal of all polynomial identities satisfied by $\mathfrak{V}$. The factor algebra $F(\mathfrak{V}) = K\langle X \rangle / T(\mathfrak{V})$ is the *relatively free algebra of countable rank in the variety* $\mathfrak{V}$. We shall use the same symbols $x_i \in X$ for the *free generators* of $F(\mathfrak{V})$. The subalgebra $F_d(\mathfrak{V})$ generated in $F(\mathfrak{V})$ by $x_1, \ldots, x_d$ is the *relatively free algebra of rank d in* $\mathfrak{V}$. We shall use the notation $F^+(\mathfrak{V})$ for the relatively free algebra of $\mathfrak{V}$ in the nonunitary case.

(ii) If R is any algebra, then the variety var R defined by all polynomial identities from $T(R)$ is called the *variety generated by the algebra R*.

(iii) The variety $\mathfrak{W}$ is called a *subvariety of* $\mathfrak{V}$ if $\mathfrak{W} \subseteq \mathfrak{V}$ as classes.

There exists a 1-1 correspondence π between the T-ideals of $K\langle X \rangle$ and the varieties of associative algebras: For every T-ideal V we define $\mathfrak{V} = \pi(V)$ to be the variety determined by the polynomial identities belonging to V. This is a "Galois correspondence", i.e., for any two T-ideals V_1, V_2 the inclusion $V_1 \subset V_2$ is equivalent to $\pi(V_1) \supset \pi(V_2)$.

Two of the main combinatorial problems on polynomial identities are the following:

Problem 1.2.5. *For a concrete algebra R find a basis for its polynomial identities.*

Problem 1.2.6. (Specht problem [Sp]) *Does every T-ideal have a finite basis?*

The first problem is especially interesting for algebras, important not only in ring theory but also in other branches of mathematics. From such a point of view, the polynomial identities of matrix algebras are among the most attractive objects to study.

Starting from the early 1960s, the Specht problem was very popular in most of the research groups working on PI-algebras. The algebras with finite bases of their polynomial identities were called *Spechtian*. If a variety $\mathfrak{V}$ as well as all its subvarieties have finite bases of identities, then $\mathfrak{V}$ satisfies the *Specht property*. Of course, the Specht problem is a partial case of the well known problem for the *finite basis property* for an arbitrary variety of abstract algebras. For example, this problem was asked by B. H. Neumann for varieties of groups in his thesis in 1935, see also his paper [Ne]. But, nevertheless, now the notion *Specht property* is used for varieties of arbitrary algebraic systems. (See [Dr10] Section 3.1, for the state of the art of the finite basis problem for various algebras.)

Over fields of characteristic 0 the Specht problem was solved into affirmative by Kemer [Ke3] in 1987, as a result of the deep structure theory of T-ideals developed by him in [Ke2], both in the unitary and nonunitary case. (See also the book [Ke5] for a self-contained exposition of the results of Kemer.) The case of associative algebras over fields of positive characteristic was open until 1999 when Belov [B4] (see also [B5]), Grishin [Gr1] and Shchigolev [Shc] constructed the first counterexamples of varieties of nonunitary associative algebras which cannot be defined by a finite system of polynomial identities (over any field of positive characteristic in the papers by Belov and Shchigolev and over a field of characteristic 2 in the paper by Grishin). Now there are a lot of new examples, see, e.g., Grishin [Gr2, Gr3], Gupta and Krasilnikov [GK1, GK2]. Maybe the simplest example is in [GK2]: Over a field of characteristic 2 the system of polynomial identities

$$[x, y^2]x_1^2 x_2^2 \cdots x_n^2 [x, y^2]^3 = 0, \quad n = 0, 1, 2, \ldots,$$

is not equivalent to a finite system. The considerations in [GK2] are in the free nonunitary algebra, and this system has the following interesting property. If we work in the class of unitary algebras, then each identity implies all previous ones and nevertheless the system is still not equivalent to any finite system.

The T-ideals have systems of generators which are especially convenient for working with.

Definition 1.2.7. The polynomial identity

$$f(x_1, \ldots, x_m) = \sum \alpha_i x_{i_1} \cdots x_{i_{d_i}} \in K\langle X \rangle, \quad \alpha_i \in K,$$

is called *homogeneous of degree d* if all monomials $x_{i_1} \cdots x_{i_{d_i}}$ with nonzero coefficients are of the same degree d; $f(x_1, \ldots, x_m)$ is *multihomogeneous of multidegree* $(d_1, \ldots, d_m)$ if each variable x_i appears the same number of times d_i in all monomials. The polynomial $f(x_1, \ldots, x_m)$ is *multilinear of degree m* if it is linear (i.e., homogeneous of first degree) in each variable $x_1, \ldots, x_m$. Hence it has the form

$$f(x_1, \ldots, x_m) = \sum_{\sigma \in S_m} \beta_\sigma x_{\sigma(1)} \cdots x_{\sigma(m)}, \quad \beta_\sigma \in K.$$

We denote by P_m the vector space of all multilinear polynomials of degree m.

Proposition 1.2.8. *Let*

$$f(x_1, \ldots, x_m) = \sum_{i=0}^{d} f_i \in K\langle X \rangle,$$

where f_i is the homogeneous component of f of degree i in x_1.

(i) *If the base field K contains more that d elements (e.g., if K is infinite), then the polynomial identities $f_i = 0$, $i = 0, 1, \ldots, d$, follow from $f = 0$.*

(ii) *If K is of characteristic 0 (or if char $K > \deg f$), then $f = 0$ is equivalent to a set of multilinear polynomial identities.*

Proof. (i) Let $U = (f)^T$ be the T-ideal of $K\langle X \rangle$ generated by f. We choose $d+1$ different elements $\xi_0, \xi_1, \ldots, \xi_d$ of K. Since U is a T-ideal,

$$f(\xi_j x_1, x_2, \ldots, x_m) = \sum_{i=0}^{d} \xi_j^i f_i(x_1, x_2, \ldots, x_m) \in U, \quad j = 0, 1, \ldots, d.$$

We consider these inclusions as a linear system modulo U, with unknowns f_i, $i = 0, 1, \ldots, d$. Since its determinant

$$\begin{vmatrix} 1 & \xi_0 & \xi_0^2 & \cdots & \xi_0^d \\ 1 & \xi_1 & \xi_1^2 & \cdots & \xi_1^d \\ 1 & \xi_2 & \xi_2^2 & \cdots & \xi_2^d \\ \vdots & \vdots & \vdots & \ddots & \vdots \\ 1 & \xi_d & \xi_d^2 & \cdots & \xi_d^d \end{vmatrix} = \prod_{i<j}(\xi_j - \xi_i)$$

is the Vandermonde determinant and is different from 0, we obtain that each $f_i(x_1, \ldots, x_m)$ also belongs to U, i.e., the polynomial identities $f_i = 0$ are consequences of $f = 0$.

(ii) We use the *process of linearization*. By (i) we may assume that f is homogeneous in each of its variables. Let $\deg_{x_1} f = d$. We write $f(y_1 + y_2, x_2, \ldots, x_m) \in U$ in the form

$$f(y_1 + y_2, x_2, \ldots, x_m) = \sum_{i=0}^{d} f_i(y_1, y_2, x_2, \ldots, x_m),$$

where f_i is the homogeneous component of degree i in y_1. Hence $f_i \in U$, $i = 0, 1, \ldots, d$. Since $\deg_{y_j} f_i < d$, $i = 1, \ldots, d-1$, $j = 1, 2$, we apply inductive arguments and obtain a set of multilinear consequences of $f = 0$. In order to see that these multilinear identities are equivalent to $f = 0$, it is sufficient to see that

$$f_i(y_1, y_1, x_2, \ldots, x_m) = \binom{d}{i} f(y_1, x_2, \ldots, x_m),$$

and the binomial coefficient is nonzero because char $K = 0$ or char $K = p > d$. $\quad\square$

Remark 1.2.9. (i) It is easy to check whether a multilinear element $f(x_1, \ldots, x_m)$ of $K\langle X \rangle$ vanishes on an algebra R: If we fix a basis $\{s_1, s_2, \ldots\}$ of R, then $f(x_1, \ldots, x_m) = 0$ is an identity for R if and only if $f(s_{i_1}, \ldots, s_{i_m}) = 0$ for all m-tuples of basis elements.

(ii) If C is any commutative algebra containing K and $|K| = \infty$, then the algebras R and $C \otimes_K R$ have the same polynomial identities. If char $K = 0$, this follows from (i): A multilinear polynomial identity $f(x_1, \ldots, x_m) = 0$ vanishes on $C \otimes_K R$ if and only if it vanishes on all m-tuples of basis elements of R. For the case of infinite fields of arbitrary characteristic one needs arguments similar to those in the proof of Proposition 1.3.2 below.

Remark 1.2.10. It is not true that over an arbitrary field K every polynomial identity is equivalent to a system of multilinear identities. Nevertheless every PI-algebra satisfies some multilinear polynomial identity. For the proof, let R be a PI-algebra and let $f(x_1, \ldots, x_m) \in K\langle X \rangle$ be a polynomial identity of degree $d > 1$ in x_1. Then

$$
\begin{aligned}
f_1(y_1, y_2, x_2, \ldots, x_m) &= f(y_1 + y_2, x_2, \ldots, x_m) \\
&\quad - f(y_1, x_2, \ldots, x_m) - f(y_2, x_2, \ldots, x_m)
\end{aligned}
$$

contains a nonzero homogeneous component of degree $d - 1$ in y_1 (which is linear in y_2) and has no homogeneous components of degree d in y_1 or y_2. Continuing the process of increasing the number of variables and decreasing their degrees, we obtain a polynomial $f_d(y_1, \ldots, y_d, x_2, \ldots, x_m)$ which is linear in $y_1, \ldots, y_d$ and, similarly, a polynomial which is linear in all variables.

1.3   Generic Matrices

Since the relatively free algebra $F(\mathfrak{V})$ of the variety of algebras $\mathfrak{V}$ is the factor algebra of the free algebra modulo the T-ideal of $\mathfrak{V}$, all the information for $\mathfrak{V}$ is encoded in $F(\mathfrak{V})$. Nevertheless, a priori it is not clear how to work in $F(\mathfrak{V})$ and how to perform concrete computations there. In the important case of polynomial identities of matrices, the relatively free algebra $F(\text{var } M_n(K))$ has a nice realization as the algebra of generic matrices, introduced by Procesi [Pr1].

We fix the integer $n \geq 2$ and denote by $\Omega = \Omega_n$ the K-algebra of the polynomials in infinitely many commuting variables

$$
\Omega_n = K[y_{pq}^{(i)} \mid p, q = 1, \ldots n, \ i = 1, 2, \ldots].
$$

Definition 1.3.1. The $n \times n$ matrices with entries from Ω_n

$$
y_i = \sum_{p,q=1}^{n} y_{pq}^{(i)} e_{pq}, \ i = 1, 2, \ldots,
$$

are called *generic $n \times n$ matrices*. The algebra R_n generated by the generic $n \times n$ matrices is the *generic $n \times n$ matrix algebra*. We denote by R_{nd} the subalgebra of R_n generated by the first d generic matrices $y_1, \ldots, y_d$.

For example, for $n = d = 2$, changing the notation to $x = y_1$, $y = y_2$ and $x_{pq} = y_{pq}^{(1)}$, $y_{pq} = y_{pq}^{(2)}$, the algebra R_{22} is generated by

$$x = \begin{pmatrix} x_{11} & x_{12} \\ x_{21} & x_{22} \end{pmatrix}, \quad y = \begin{pmatrix} y_{11} & y_{12} \\ y_{21} & y_{22} \end{pmatrix}.$$

For any commutative K-algebra C, the $n \times n$ matrices with entries from C can be obtained by specializations of the generic matrices, e.g.,

$$a = \sum_{p,q=1}^{n} \gamma_{pq} e_{pq}, \gamma_{pq} \in C,$$

is obtained from

$$y_1 = \sum_{p,q=1}^{n} y_{pq}^{(1)} e_{pq}$$

by replacing the variables $y_{pq}^{(1)}$ with γ_{pq}.

Proposition 1.3.2. *If the base field K is infinite, then the generic matrix algebra R_n is isomorphic to the relatively free algebra $F(\mathrm{var}\, M_n(K))$ of the variety generated by the $n \times n$ matrix algebra. If K is a finite field and P is any infinite extension of K, then $R_n \cong F(\mathrm{var}\, M_n(P))$, (where $M_n(P)$ is considered as a K-algebra).*

Proof. Let P be any infinite field containing K and let us consider the canonical homomorphisms

$$\pi : K\langle X \rangle \longrightarrow F(M_n(P)), \quad \rho : K\langle X \rangle \longrightarrow R_n.$$

We shall show that the kernels of π and ρ coincide, i.e., $f(x_1, \ldots, x_m) \in K\langle X \rangle$ is a polynomial identity for $M_n(P)$ if and only if $f(y_1, \ldots, y_m) = 0$ in R_n. Clearly, if $f(x_1, \ldots, x_m) \in \ker \rho$, then $f(y_1, \ldots, y_m) = 0$. For any matrices $r_1, \ldots, r_m$ in $M_n(P)$, there is a natural homomorphism $R_n \to M_n(P)$ defined by $y_i \to r_i$ for $i = 1, \ldots, m$, and $y_i \to 0$ for $i > m$. Hence $f(r_1, \ldots, r_m) = 0$ and $f(x_1, \ldots, x_m) = 0$ is a polynomial identity for $M_n(P)$. Now, let $f(x_1, \ldots, x_m) \in \ker \pi$, i.e., $f(x_1, \ldots, x_m) = 0$ is an identity for $M_n(P)$, and let $f(y_1, \ldots, y_m) \neq 0$ in R_n. The entries f_{pq} of $f(y_1, \ldots, y_m)$ are polynomials in the commuting variables $y_{pq}^{(i)}$. Since P is an infinite field containing K and some $f_{p_0 q_0} = f_{p_0 q_0}(y_{pq}^{(i)})$ is a nonzero polynomial in the $y_{pq}^{(i)}$'s, we can find elements $\xi_{pq}^{(i)} \in P$ such that $f_{p_0 q_0}(\xi_{pq}^{(i)}) \neq 0$. This means that for the matrices

$$r_i = \sum_{p,q=1}^{n} \xi_{pq}^{(i)} e_{pq}, \quad i = 1, \ldots, m,$$

the expression $f(r_1, \ldots, r_m)$ is different from zero, which is impossible. Hence $f(y_1, \ldots, y_m) = 0$ and $\ker \rho = \ker \pi$.   □

Lemma 1.3.3. *The eigenvalues of the $n \times n$ generic matrix y_1 are pairwise different.*

Proof. We consider y_1 as a matrix with entries from the field of fractions of the polynomial algebra Ω_n. Let $f(\lambda)$ be the characteristic polynomial of y_1 and let $f(\lambda)$ have multiple zeros. Then the discriminant of $f(\lambda)$ is equal to zero. Since every $n \times n$ matrix a with entries from K is a specialization of y_1, the discriminant of the characteristic polynomial $f_a(\lambda)$ is also equal to 0 and this means that a has multiple eigenvalues. This is impossible because there exists a matrix $a \in M_n(K)$ without multiple eigenvalues. (This is trivial if the field K is infinite and a little bit tricky if $K = \mathbb{F}_q$, see, e.g., [Dr10], Exercises 7.2.3, 7.2.4 and Lemma 7.2.5.)   □

By the following corollary we may assume that one of the generic matrices is diagonal.

Corollary 1.3.4. *Let Ω_n be as above and let*

$$y_1' = \sum_{p=1}^{n} y_{pp}^{(1)} e_{pp}, \quad y_i' = y_i, \quad i > 1.$$

The algebra R_n' generated by $y_1', y_2', y_3', \ldots$ is isomorphic to the generic matrix algebra R_n.

Proof. Let Ξ be the algebraic closure of the field of fractions of the polynomial algebra Ω_n. By Lemma 1.3.3, the generic matrix y_1 has no multiple eigenvalues. Hence there exists an invertible matrix z with entries from Ξ such that the matrix $u_1 = z^{-1} y_1 z$ is diagonal. Let

$$u_i = z^{-1} y_i z, \quad i = 1, 2, \ldots.$$

Denote by U_n the K-subalgebra of $M_n(\Xi)$ generated by $u_1, u_2, \ldots$. Since the matrices y_i' are obtained as specializations of the generic matrices y_i, there is a homomorphism $\varphi : R_n \to R_n'$ defined by $\varphi(y_i) = y_i'$, $i = 1, 2, \ldots$. Similarly, the matrices u_i are specializations of y_i' and again there is a homomorphism ψ which sends y_i' to u_i, $i = 1, 2, \ldots$. The composition $\psi \circ \varphi : R_n \longrightarrow U_n$ is the isomorphism defined by $y_i \to u_i = z^{-1} y_i z$. This implies that $\ker \varphi = 0$ and, since φ is onto R_n', it is an isomorphism.   □

Remark 1.3.5. We may assume even more, that the first generic matrix in R_n is diagonal and the second has the same first row and column, e.g.,

$$y_1 = \sum_{p=1}^{n} y_{pp}^{(1)} e_{pp}, \quad y_2 = \sum_{p,q=1}^{n} y_{pq}^{(2)} e_{pq},$$

and $y_{1q}^{(2)} = y_{q1}^{(2)}$, $q = 2, 3, \ldots, n$.

Remark 1.3.6. Let S be a finite dimensional algebra with basis $\{s_1, \ldots, s_m\}$ over an infinite field K and let

$$\Omega_S = K[z_p^{(i)} \mid p = 1, \ldots, m, \ i = 1, 2, \ldots]$$

be the polynomial algebra. Let us consider the "generic" algebra R_S generated as a K-algebra by the elements

$$z_i = \sum_{p=1}^{m} z_p^{(i)} s_p, \quad i = 1, 2, \ldots,$$

where we assume that the polynomials of Ω_S commute with the basis elements of S and the multiplication between $s_1, \ldots, s_m$ is as in S. (In other words, R_S is naturally embedded into the tensor product of K-algebras $\Omega_S \otimes S$.)

Repeating the arguments from Proposition 1.3.2 we can show that the algebra R_S is isomorphic to the relatively free algebra $F(\mathrm{var}\, S)$.

Kemer [Ke4] showed that the polynomial identities of any finitely generated algebra over a field of characteristic 0 coincide with the polynomial identities of some finite dimensional algebra. Hence, if $\mathfrak{V}$ is a variety of algebras and $F_d(\mathfrak{V})$ is the relatively free algebra of finite rank d in $\mathfrak{V}$, then there exists a finite dimensional algebra S such that $T(F_d(\mathfrak{V})) = T(S)$. If $R_{S,d}$ is the subalgebra of R_S generated by $z_1, \ldots, z_d$, then the algebras $F_d(\mathfrak{V})$ and $R_{S,d}$ are isomorphic.

1.4 Polynomial Identities of the Grassmann Algebra

The complete description of the polynomial identities of concrete algebras is known in very few cases only. In this section we shall give a basis of the polynomial identities of the Grassmann algebra. This is a theorem of Krakowski and Regev [KR] established in 1973. (The original theorem of Krakowski and Regev is stronger and gives additional information on the polynomial identities, e.g., the cocharacter sequence of E.) First, as an easy exercise, we shall describe the polynomial identities in commutative algebras.

Lemma 1.4.1. *Let the algebra R be commutative. Then its T-ideal is $T(R) = ([x_1, x_2])^T$. Working with a nonunitary algebra R, and depending on whether R is not nilpotent or nilpotent, its T-ideal in $K^+\langle X \rangle$ is*

$$T(R) = ([x_1, x_2])^T \quad or \quad T(R) = ([x_1, x_2], x_1 \cdots x_m)^T.$$

Proof. We shall consider the nonunitary case only. The algebra R satisfies the polynomial identity $[x_1, x_2] = 0$ and we shall work modulo it. The field is of characteristic 0 and we may assume that

$$f(x_1, \ldots, x_m) = \sum_{\sigma \in S_m} \beta_\sigma x_{\sigma(1)} \cdots x_{\sigma(m)} = 0, \quad \beta_\sigma \in K,$$

is a multilinear polynomial identity for R. Modulo $[x_1, x_2] = 0$ we can rearrange the monomials $\beta_\sigma x_{\sigma(1)} \cdots x_{\sigma(m)}$ in the form $\beta_\sigma x_1 \cdots x_m$ and present f in the form

$$f(x_1, \ldots, x_m) = \beta x_1 \cdots x_m, \quad \beta \in K.$$

If $\beta = 0$, then $f(x_1, \ldots, x_m) = 0$ is a consequence of the identity $[x_1, x_2] = 0$. If $\beta \neq 0$, then, modulo $[x_1, x_2] = 0$, the identity $f(x_1, \ldots, x_m) = 0$ is equivalent to the identity of nilpotency $x_1 \cdots x_m = 0$. Choosing the minimal m such that $x_1 \cdots x_m = 0$ is an identity for R, we complete the proof.   □

As we already know, see Example 1.1.6, the Grassmann algebra satisfies the identity $[x_1, x_2, x_3] = 0$.

Lemma 1.4.2. *The polynomial identity*

$$[x_1, x_2][x_3, x_4] + [x_1, x_3][x_2, x_4] = 0$$

is a consequence of the identity $[x_1, x_2, x_3] = 0$.

Proof. We shall work modulo the identity $[x_1, x_2, x_3] = 0$. We apply the identities

$$[uv, w] = [u, w]v + u[v, w], \quad [w, uv] = [w, u]v + u[w, v]$$

(which express the fact that the commuting with a fixed element is a derivation) and obtain from $[x_1, x_2 x_3, x_4] = 0$ that

$$[x_1, x_2 x_3, x_4] = [[x_1, x_2]x_3 + x_2[x_1, x_3], x_4]$$
$$= [x_1, x_2, x_4]x_3 + [x_1, x_2][x_3, x_4] + [x_2, x_4][x_1, x_3] + x_2[x_1, x_3, x_4] = 0,$$

$$[x_1, x_2][x_3, x_4] + [x_2, x_4][x_1, x_3] = 0.$$

Since $[x_1, x_2, x_3] = [x_1, x_2]x_3 - x_3[x_1, x_2] = 0$, we have that the commutators $[x_i, x_j]$ commute with all elements modulo $[x_1, x_2, x_3] = 0$. Hence $[x_2, x_4][x_1, x_3] = [x_1, x_3][x_2, x_4]$ and we obtain the identity

$$[x_1, x_2][x_3, x_4] + [x_1, x_3][x_2, x_4] = 0.$$ □

Lemma 1.4.3. *Modulo the polynomial identity $[x_1, x_2, x_3] = 0$, every multilinear polynomial can be presented in the form*

$$f(x_1, \ldots, x_m) = \sum \beta_{i,j} x_{i_1} \cdots x_{i_{m-2p}}[x_{j_1}, x_{j_2}] \cdots [x_{j_{2p-1}}, x_{j_{2p}}], \quad \beta_{i,j} \in K,$$

where the sum runs on all permutations $[i_1, \ldots, i_{m-2p}, j_1, j_2, \ldots, j_{2p-1}, j_{2p}]$ of $1, \ldots, m$ such that $i_1 < \cdots < i_{m-2p}$, $j_1 < j_2 < \cdots < j_{2p-1} < j_{2p}$.

Proof. We shall work modulo the identity $[x_1, x_2, x_3] = 0$. Since

$$x_2 x_1 = x_1 x_2 - [x_1, x_2]$$

and the commutators commute with all elements, we obtain for $j > i$ the identity

$$x_j x_i x_{k_1} \cdots x_{k_{m-2}} = (x_i x_j - [x_i, x_j]) x_{k_1} \cdots x_{k_{m-2}}$$
$$= x_i x_j x_{k_1} \cdots x_{k_{m-2}} - x_{k_1} \cdots x_{k_{m-2}} [x_i, x_j].$$

Hence, we may present the multilinear polynomial $f(x_1, \ldots, x_m)$ in the form

$$f(x_1, \ldots, x_m) = \sum \gamma_{i,j} x_{i_1} \cdots x_{i_{m-2p}} [x_{j_1}, x_{j_2}] \cdots [x_{j_{2q-1}}, x_{j_{2p}}],$$

where $\gamma_{i,j} \in K$ and $i_1 < \cdots < i_{m-2p}$. Using the identity $[x_1, x_2][x_3, x_4] + [x_1, x_3][x_2, x_4] = 0$ and the trivial identity $[x_2, x_1] = -[x_1, x_2]$, we obtain that

$$[x_{\sigma(1)}, x_{\sigma(2)}] \cdots [x_{\sigma(2p-1)}, x_{\sigma(2p)}] = (\text{sign } \sigma)[x_1, x_2] \cdots [x_{2p-1}, x_{2p}]$$

and this allows to rearrange also the indices $j_1, j_2, \ldots, j_{2p-1}, j_{2p}$ in the product of commutators. $\square$

Theorem 1.4.4. *The polynomial identity*

$$[x_1, x_2, x_3] = 0$$

forms a basis of the polynomial identities of the Grassmann algebra E of an infinite dimensional vector space.

Proof. By Example 1.1.6, the algebra E satisfies the identity $[x_1, x_2, x_3] = 0$. Let $f(x_1, \ldots, x_m) = 0$ be a polynomial identity of E which does not follow from the commutator of length 3. We shall prove the theorem if we establish that this is impossible and $f(x_1, \ldots, x_m)$ vanishes modulo the identity $[x_1, x_2, x_3] = 0$.

Let $f(x_1, \ldots, x_m)$ be not zero modulo $[x_1, x_2, x_3] = 0$. By Lemma 1.4.3, $f(x_1, \ldots, x_m)$ has the form

$$f(x_1, \ldots, x_m) = \sum \beta_{i,j} x_{i_1} \cdots x_{i_{m-2p}} [x_{j_1}, x_{j_2}] \cdots [x_{j_{2p-1}}, x_{j_{2p}}],$$

where $\beta_{i,j} \in K$ and $i_1 < \cdots < i_{m-2p}$, $j_1 < j_2 < \cdots < j_{2p-1} < j_{2p}$. We choose the minimal p with $\beta_{i,j} \neq 0$ and fix one of the corresponding $(m - 2p, 2p)$-tuples $(i_1, \ldots, i_{m-2p}; j_1, j_2, \ldots, j_{2p-1}, j_{2p})$. We replace $x_{i_1}, \ldots, x_{i_{m-2p}}$ respectively with the products of two generators of the Grassmann algebra $a_{i_1} = e_1 e_2, \ldots, a_{i_{m-2p}} = e_{2m-4p-1} e_{2m-4p}$ and $x_{j_1}, x_{j_2}, \ldots, x_{j_{2p-1}}, x_{j_{2p}}$ respectively with the generators $a_{j_1} = e_{2m-4p+1}, a_{j_2} = e_{2m-4p+2}, \ldots, a_{j_{2p-1}} = e_{2m-2p-1}, a_{j_{2p}} = e_{2m-2p}$. Clearly,

$$a_{i_1} \cdots a_{i_{m-2p}} [a_{j_1}, a_{j_2}] \cdots [a_{j_{2p-1}}, a_{j_{2p}}]$$
$$= 2^p e_1 e_2 \cdots e_{2m-4p-1} e_{2m-4p} e_{2m-4p+1} e_{2m-4p+2} \cdots e_{2m-2p-1} e_{2m-2p} \neq 0.$$

Since all products $e_a e_b$ are in the centre of E and $[e_a e_b, g] = 0$ for $g \in E$, the minimality of p gives that all other summands $a_{i'_1} \cdots a_{i'_{m-2p'}} [a_{j'_1}, a_{j'_2}] \cdots [a_{j'_{2p'-1}}, a_{j'_{2p'}}]$ vanish and $f(x_1, \ldots, x_m) = 0$ is not a polynomial identity for the Grassmann algebra. (The same arguments work if we replace $x_{j_1}, x_{j_2}, \ldots, x_{j_{2p-1}}, x_{j_{2p}}$ with $1 \in E$.) $\square$

Remark 1.4.5. Considerations similar to those of the proof of Theorem 1.4.4 show that the polynomial identities of the Grassmann algebra E_n of the n-dimensional vector space V_n, $n > 1$, follow from $[x_1, x_2, x_3] = 0$ and the standard identity $s_{2k}(x_1, \ldots, x_{2k}) = 0$, where k is the minimal integer with $2k > n$.

As we already said at the beginning of this section, the explicit polynomial identities for concrete algebras are known in very few cases. Among them are the algebra $UT_n(K)$ of $n \times n$ upper triangular matrices, the algebra $M_2(K)$ of all 2×2 matrices and the tensor square $E \otimes E$ of the Grassmann algebra. Yu. N. Maltsev [Mts] showed in 1971 that the basis of the identities of $UT_n(K)$ consists of the single identity $[x_1, x_2] \cdots [x_{2n-1}, x_{2n}] = 0$. Razmyslov [Ra1] in 1973 found a basis of 9 identities of degree 4, 5 and 6 for $M_2(K)$ and Drensky [Dr2] in 1981 showed that the identity $s_4(x_1, x_2, x_3, x_4) = 0$ and the specialization of the Hall identity $[[x_1, x_2]^2, x_1] = 0$ form a minimal basis of the identities of $M_2(K)$. Popov [Po] proved that the basis of the identities of $E \otimes E$ consists of the centre-by-metabelian identity $[[[x_1, x_2], [x_3, x_4]], x_5] = 0$ and the identity $[[x_1, x_2]^2, x_1] = 0$. See also the comments in [Dr10] (the beginning of Chapter 5, p. 49) for related results over fields of positive characteristic.

Chapter 2

Quantitative Approach to PI-algebras

2.1 Codimensions and Hilbert Series

By Proposition 1.2.8, if the base field K is of characteristic 0, then every polynomial identity $f(x_1, \ldots, x_m) = 0$ is equivalent to a collection of multilinear identities. Also, see Remark 1.2.9 (i), it is very easy to check whether a multilinear identity is satisfied by an algebra R. The multilinear identities are interesting also in the case of arbitrary field because by Remark 1.2.10 any PI-algebra satisfies a multilinear polynomial identity. One may use the sequence of dimensions of the multilinear polynomial identities of an algebra R (or in a T-ideal) as a measure of the polynomial identities of R. By the theorem of Regev [Re1] (see the last chapter of these lecture notes) it turned out that it is more convenient to study the sequence of codimensions.

Definition 2.1.1. Let P_m be the vector subspace of all multilinear elements of degree m in the free algebra $K\langle X \rangle$ and let R be an algebra with T-ideal $T(R)$. The dimension of the multilinear polynomials in $K\langle X \rangle$ modulo the polynomial identities of R is called the *m-th codimension of the T-ideal $T(R)$ or of the polynomial identities of R* and is denoted by $c_m(R)$ (or by $c_m(\mathfrak{V})$ if we consider the T-ideal of the polynomial identities of a variety $\mathfrak{V}$), i.e.,

$$c_m(R) = \dim P_m/(P_m \cap T(R)), \ m = 0, 1, 2, \ldots.$$

This sequence is called the *codimension sequence* of the T-ideal $T(R)$.

Example 2.1.2. By Theorem 1.4.4 all polynomial identities of the Grassmann algebra of an infinite dimensional vector space are consequences of the identity $[x_1, x_2, x_3] = 0$. We have seen in the proof of this theorem that the vector space $P_m/(P_m \cap T(E))$ has a basis consisting of all multilinear elements of the form

$$x_{i_1} \cdots x_{i_{m-2p}} [x_{j_1}, x_{j_2}] \cdots [x_{j_{2p-1}}, x_{j_{2p}}],$$

where $i_1 < \cdots < i_{m-2p}$ and $j_1 < j_2 < \cdots < j_{2p-1} < j_{2p}$. Hence, if $m > 0$, then

$$c_m(E) = \binom{m}{0} + \binom{m}{2} + \binom{m}{4} + \cdots = 2^{m-1}$$

(and $c_0(E) = 1$ or $c_0(E) = 0$ if we consider E respectively as a unitary or a nonunitary algebra).

Example 2.1.3. It is a little bit more complicated to show that the factor space $P_m/(P_m \cap T(UT_2(K)))$ associated with the polynomial identities of the algebra of 2×2 upper triangular matrices has a basis consisting of multilinear elements of the form

$$x_1 \cdots x_m, \quad x_{i_1} \cdots x_{i_{m-p}}[x_{j_1}, x_{j_2}, \ldots, x_{j_p}],$$

where $i_1 < \cdots < i_{m-p}$, $j_1 > j_2 < \cdots < j_p$, for $p = 2, 3, \ldots, m$. A simple counting gives that

$$c_m(UT_2(K)) = 1 + \sum_{p=2}^{m} \binom{m}{p}(m-1) = 2^{m-1}(m-2) + 2, \quad m \geq 1.$$

In the last chapter we shall prove the theorem of Regev [Re1] stating that the codimension sequence of any PI-algebra is exponentially bounded. As a consequence of this theorem we shall obtain that the tensor product of two PI-algebras is PI again. We shall also discuss there other results on the growth of codimensions of PI-algebras.

The T-ideal of $UT_2(K)$ is generated by the product $[x_1, x_2][x_3, x_4]$ of two commutators and is equal to the product of two copies of the commutator ideal of $K\langle X \rangle$ (the T-ideal generated by the commutator $[x_1, x_2]$). In the end of the section we shall see how to express the codimensions and other numerical invariants of a T-ideal W which is a product of two T-ideals U and V in terms of the numerical invariants of U and V.

Every polynomial identity can be written in the form

$$f(x_1, \ldots, x_m) = \sum f_{(a_1, \ldots, a_m)} = 0,$$

where $f_{(a_1, \ldots, a_m)}$ is the multihomogeneous component of f which is of degree a_i in x_i. By Proposition 1.2.8, if the base field K is infinite, then $f = 0$ is equivalent to its multihomogeneous components $f_{(a_1, \ldots, a_m)} = 0$. Hence every T-ideal U of $K\langle X \rangle$ is a homogeneous ideal, i.e., is a direct sum of its multihomogeneous components. If we consider a (multi)graded vector space such that all homogeneous components are finite dimensional, it is convenient to use its Hilbert series to measure it.

Definition 2.1.4. (i) The vector space V is *graded* if it is a direct sum of its subspaces V_m, $m \geq 0$, i.e.,

$$V = V_0 \oplus V_1 \oplus V_2 \oplus \cdots.$$

The subspaces V_m are called the *homogeneous components of degree m of V*. Similarly, V is *multigraded* if

$$V = \bigoplus_{m_i \geq 0} V_{(m_1, \ldots, m_d)},$$

where $V_{(m_1, \ldots, m_d)}$ is its *homogeneous component of degree* $(m_1, \ldots, m_d)$ and the direct sum runs on all d-tuples $(m_1, \ldots, m_d)$ with $m_i \geq 0$. The subspace W of the graded vector space $V = \bigoplus_{m \geq 0} V_m$ is a *graded* or (*homogeneous*) *subspace* if $W = \bigoplus_{m \geq 0}(W \cap V_m)$. In this case, the factor space V/W can also be naturally graded and *V/W inherits the grading of V*.

(ii) If $V = \bigoplus_{m \geq 0} V_m$ is a graded vector space and dim $V_m < \infty$ for all $m \geq 0$, then the formal power series

$$H(V, t) = \sum_{m \geq 0} (\dim V_m) t^m$$

is called the *Hilbert* (or *Poincaré*) *series* of V. If the vector space

$$V = \bigoplus_m V_{(m_1, \ldots, m_d)}, \quad m = (m_1, \ldots, m_d),$$

is multigraded, then the *Hilbert series* of V is

$$H(V, t_1, \ldots, t_d) = \sum_m (\dim V_{(m_1, \ldots, m_d)}) t_1^{m_1} \cdots t_d^{m_d}.$$

For example, the polynomial algebra $K[x_1, \ldots, x_d]$ is graded assuming that the homogeneous polynomials of degree m (in the usual sense) are the homogeneous elements of degree m. Similarly, $K[x_1, \ldots, x_d]$ has a multigrading, counting the entry of each variable in the monomials. Analogously, one can define grading and multigrading on the free associative algebra $K\langle x_1, \ldots, x_d \rangle$ of finite rank d and a grading on $K\langle X \rangle$. Usually we shall assume that $K[x_1, \ldots, x_d]$ and $K\langle x_1, \ldots, x_d \rangle$ are equipped with these two gradings. Their Hilbert series are

$$H(K[x_1, \ldots, x_d], t_1, \ldots, t_d) = \prod_{i=1}^{d} \frac{1}{1 - t_i}, \quad H(K[x_1, \ldots, x_d], t) = \frac{1}{(1-t)^d},$$

$$H(K\langle x_1, \ldots, x_d \rangle, t_1, \ldots, t_d) = \frac{1}{1 - (t_1 + \cdots + t_d)},$$

$$H(K\langle x_1, \ldots, x_d \rangle, t) = \frac{1}{1 - dt}.$$

If $V = \bigoplus V_m$ and $W = \bigoplus W_m$ are graded vector spaces with the same (multi)grading, then $V \oplus W$ and $V \otimes W$ are also graded with homogeneous components

$$(V \oplus W)_m = V_m \oplus W_m, \quad (V \otimes W)_m = \bigoplus_{m' + m'' = m} V_{m'} \otimes W_{m''}.$$

The Hilbert series of $V \oplus W$, $V \otimes W$ and of the factor space V/U, where $U \subseteq V$, satisfy the relations

$$
\begin{aligned}
H(V \oplus W, t) &= H(V, t) + H(W, t), \\
H(V \otimes W, t) &= H(V, t) \cdot H(W, t), \\
H(V, t) &= H(V/U, t) + H(U, t).
\end{aligned}
$$

Example 2.1.5. Working with multihomogeneous elements (instead of with multi-linear elements only) and repeating the main steps of the proof of Theorem 1.4.4, one can see that the algebra $F(\text{var } E) = K\langle X \rangle / T(E)$ has a basis consisting of all elements

$$x_{i_1} \cdots x_{i_{m-2p}} [x_{j_1}, x_{j_2}] \cdots [x_{j_{2p-1}}, x_{j_{2p}}],$$

where $i_1 \leq \cdots \leq i_{m-2p}$ and $j_1 < j_2 < \cdots < j_{2p-1} < j_{2p}$. Hence, if $d > 1$, then the Hilbert series of the relatively free algebra $F_d(\text{var } E)$ is

$$H(F_d(\text{var } E), t_1, \ldots, t_d) = \left(\sum_{m_i \geq 0} t_1^{m_1} \cdots t_d^{m_d} \right) \left(\sum_{j_1 < \cdots < j_{2p}} t_{j_1} \cdots t_{j_{2p}} \right)$$

$$= \frac{1}{2} \left(\prod_{i=1}^{d} (1 - t_i) + \prod_{i=1}^{d} (1 + t_i) \right) \prod_{i=1}^{d} \frac{1}{1 - t_i} = \frac{1}{2} + \frac{1}{2} \prod_{i=1}^{d} \frac{1 + t_i}{1 - t_i}.$$

See also Carini [C] for an alternative approach.

In his survey article [F4] Formanek gave a formula for the Hilbert series of the product of two T-ideals as a function of the Hilbert series of the factors. The proof of the result is given in the paper by Halpin [Ha2]. It works for arbitrary homogeneous ideals of the free algebra.

Proposition 2.1.6. (Formanek [F4], see also Halpin [Ha2]) *Let U and V be mul-tihomogeneous ideals of the free algebra $K\langle x_1, \ldots, x_d \rangle$. Then the Hilbert series of UV, U and V are related by the equation*

$$H(U, t_1, \ldots, t_d) H(V, t_1, \ldots, t_d)$$
$$= H(UV, t_1, \ldots, t_d) H(K\langle x_1, \ldots, x_d \rangle, t_1, \ldots, t_d).$$

Proof. It is well known, see Cohn [Co], that the free algebra $A_d = K\langle x_1, \ldots, x_d \rangle$ is a FIR (free ideal ring) and every one-sided ideal of A_d is a free A_d-module. If the ideal is homogeneous, then it has a homogeneous system of free generators. Similar multihomogeneous systems exist also in the multihomogeneous case. Hence there exist two systems of multihomogeneous polynomials $u_1, u_2, \ldots$ and $v_1, v_2, \ldots$ such that

$$U = \bigoplus_i u_i A_d, \quad V = \bigoplus_j A_d v_j.$$

Hence

$$H(U, t_1, \ldots, t_d) = \sum t^{|u_i|} H(A_d, t_1, \ldots, t_d),$$

$$H(V, t_1, \ldots, t_d) = \sum t^{|v_j|} H(A_d, t_1, \ldots, t_d),$$

where for a multihomogeneous u of degree p_k in x_k we denote $t^{|u|} = t_1^{p_1} \cdots t_d^{p_d}$. Using that $A_d A_d = A_d$ we obtain

$$UV = \bigoplus_{i,j} u_i A_d A_d v_j = \bigoplus_{i,j} u_i A_d v_j,$$

$$H(UV, t_1, \ldots, t_d) = \sum t^{|u_i|} t^{|v_j|} H(A_d, t_1, \ldots, t_d)$$
$$= \frac{H(U, t_1, \ldots, t_d) H(V, t_1, \ldots, t_d)}{H(A_d, t_1, \ldots, t_d)},$$

which completes the proof. □

Corollary 2.1.7. (See [F4, Ha2] or, e.g., [Dr5], Proposition 2.1) *Let* $\mathfrak{U}$, $\mathfrak{V}$ *and* $\mathfrak{W}$ *be varieties of algebras over an infinite field* K *such that* $T(\mathfrak{W}) = T(\mathfrak{U})T(\mathfrak{V})$. *Then the Hilbert series of the relatively free algebras in* $\mathfrak{U}$, $\mathfrak{V}$ *and* $\mathfrak{W}$ *satisfy the equation*

$$H(F_d(\mathfrak{W}), t_1, \ldots, t_d) = H(F_d(\mathfrak{U}), t_1, \ldots, t_d) + H(F_d(\mathfrak{V}), t_1, \ldots, t_d)$$
$$+ (t_1 + \cdots + t_d - 1) H(F_d(\mathfrak{U}), t_1, \ldots, t_d) H(F_d(\mathfrak{V}), t_1, \ldots, t_d).$$

Proof. For every variety $\mathfrak{M}$

$$H(F_d(\mathfrak{M}), t_1, \ldots, t_d) + H(A_d \cap T(\mathfrak{M}), t_1, \ldots, t_d) = H(A_d, t_1, \ldots, t_d)$$
$$= \frac{1}{1 - (t_1 + \cdots + t_d)}.$$

The equality in the statement of the corollary is obtained by the expression of the Hilbert series of $A_d \cap T(\mathfrak{M})$ in terms of the Hilbert series of $F_d(\mathfrak{M})$ for $\mathfrak{M} = \mathfrak{U}, \mathfrak{V}, \mathfrak{W}$, and then by replacing them in the equation in Proposition 2.1.6. □

Example 2.1.8. Let $\mathfrak{U}_2$ be the variety generated by the algebra $UT_2(K)$ of 2×2 upper triangular matrices over an infinite field. Its T-ideal is equal to the square of the T-ideal $T(\mathfrak{U}) = T(K)$ of the abelian variety $\mathfrak{U}$ consisting of all commutative algebras. Since $T(K)$ coincides with the commutator ideal generated as a T-ideal by the commutator $[x_1, x_2]$, we have $F_d(\mathfrak{U}) = K[x_1, \ldots, x_d]$ and

$$H(F_d(\mathfrak{U}), t_1, \ldots, t_d) = \prod_{i=1}^{d} \frac{1}{1 - t_i}.$$

Then Corollary 2.1.7 gives

$$H(F_d(\mathfrak{U}_2), t_1, \ldots, t_d) = 2H(F_d(\mathfrak{U}), t_1, \ldots, t_d)$$
$$+ (t_1 + \cdots + t_d - 1) H^2(F_d(\mathfrak{U}), t_1, \ldots, t_d)$$
$$= 2 \prod_{i=1}^{d} \frac{1}{1 - t_i} + (t_1 + \cdots + t_d - 1) \prod_{i=1}^{d} \frac{1}{(1 - t_i)^2}.$$

Any sequence $a_0, a_1, a_2, \ldots$ is encoded with its *generating function* and its *exponential generating function*. They are equal, respectively, to the formal power series

$$a(t) = \sum_{m \geq 0} a_m t^m, \quad \tilde{a}(t) = \sum_{m \geq 0} a_m \frac{t^m}{m!}.$$

We introduce the corresponding notions for the codimension sequence of a PI-algebra R. The *codimension series* and the *exponential codimension series* of R are, respectively,

$$c(R,t) = \sum_{m \geq 0} c_m(R)t^m, \quad \tilde{c}(R,t) = \sum_{m \geq 0} c_m(R)\frac{t^m}{m!}.$$

We shall write $c(\mathfrak{V},t)$ and $\tilde{c}(\mathfrak{V},t)$ when we consider codimensions of a variety of algebras $\mathfrak{V}$. The exponential codimension series of a variety $\mathfrak{V}$ is known also as the *complexity function* of $\mathfrak{V}$, see the book by Razmyslov [Ra5]. It appeared in his work on varieties of Lie algebras, the work of the author in the study of products of some T-ideals [Dr5] and in several papers by Petrogradsky, see, e.g., [Pe1, Pe2] in his classification of the growth type of varieties of Lie algebras. In the recent paper [Pe3] Petrogradsky started the systematic study of exponential codimension series of varieties of associative algebras over a field of any characteristic. The next theorem [Dr5, Pe3] describes the codimensions of the product of two T-ideals in terms of exponential codimension series. Another version of the same result is given by Berele and Regev [BR2].

Theorem 2.1.9. ([Dr5], Corollary 2.3, [Pe3], Lemma 2.1) *Let $\mathfrak{U}$, $\mathfrak{V}$ and $\mathfrak{W}$ be varieties of algebras over an infinite field such that $T(\mathfrak{W}) = T(\mathfrak{U})T(\mathfrak{V})$. Then the exponential codimension series of $\mathfrak{U}$, $\mathfrak{V}$ and $\mathfrak{W}$ satisfy the equation*

$$\tilde{c}(\mathfrak{W},t) = \tilde{c}(\mathfrak{U},t) + \tilde{c}(\mathfrak{V},t) + (t-1)\tilde{c}(\mathfrak{U},t)\tilde{c}(\mathfrak{V},t).$$

Proof. Since $c_m(\mathfrak{W})$ is equal to the coefficient of $t_1 \cdots t_m$ in $H(F_d(\mathfrak{W}), t_1, \ldots, t_d)$, $d \geq m$, by Corollary 2.1.7 we obtain

$$c_m(\mathfrak{W}) = c_m(\mathfrak{U}) + c_m(\mathfrak{V})$$
$$+ \sum_{k=0}^{m-1} \frac{m!}{k!(m-k-1)!}c_k(\mathfrak{U})c_{m-k-1}(\mathfrak{V}) - \sum_{k=0}^{m} \frac{m!}{k!(m-k)!}c_k(\mathfrak{U})c_{m-k}(\mathfrak{V}),$$

$$\sum_{m \geq 0} c_m(\mathfrak{W})\frac{t^m}{m!} = \sum_{m \geq 0} c_m(\mathfrak{U})\frac{t^m}{m!} + \sum_{m \geq 0} c_m(\mathfrak{V})\frac{t^m}{m!}$$
$$+ t\sum_{p \geq 0} c_p(\mathfrak{U})\frac{t^p}{p!} \sum_{q \geq 0} c_q(\mathfrak{V})\frac{t^q}{q!} - \sum_{p \geq 0} c_p(\mathfrak{U})\frac{t^p}{p!} \sum_{q \geq 0} c_q(\mathfrak{V})\frac{t^q}{q!},$$

which gives the relation between the exponential codimension series. $\square$

Corollary 2.1.10. (i) (Petrogradsky [Pe3], Corollary 2.2) *Let $\mathfrak{U}$ and $\mathfrak{W}_k$ be varieties of algebras with T-ideals U and $W = U^k$, respectively. Then the exponential codimension series of $\mathfrak{U}$ and $\mathfrak{W}_k$ are related by*

$$\tilde{c}(\mathfrak{W}_k,t) = \frac{(1+(t-1)\tilde{c}(\mathfrak{U},t))^k - 1}{t-1}.$$

(ii) (Drensky [Dr5], Corollary 2.4, Petrogradsky [Pe3], Theorem 3.1) *The exponential codimension series of the algebra $UT_k(K)$ of $k \times k$ upper triangular matrices is*

$$\tilde{c}(UT_k(K), t) = \frac{(1 + (t-1)e^t)^k - 1}{t - 1}.$$

Proof. (i) Let us denote $\tilde{c}(\mathfrak{U}, t)$ by $\tilde{c}(t)$. We apply Theorem 2.1.9 and induction on k:

$$\tilde{c}(\mathfrak{W}_2, t) = 2\tilde{c}(t) + (t-1)\tilde{c}^2(t)$$
$$= \frac{(1 + 2(t-1)\tilde{c}(t) + (t-1)^2\tilde{c}^2(t)) - 1}{t-1} = \frac{(1 + (t-1)\tilde{c}(t))^2 - 1}{t-1},$$

$$\tilde{c}(\mathfrak{W}_k, t) = \tilde{c}(\mathfrak{W}_{k-1}, t) + \tilde{c}(\mathfrak{U}, t) + (t-1)\tilde{c}(\mathfrak{W}_{k-1}, t)\tilde{c}(\mathfrak{U}, t)$$
$$= \frac{(1 + (t-1)\tilde{c}(t))^{k-1} - 1}{t-1} + \tilde{c}(t) + \frac{(1 + (t-1)\tilde{c}(t))^{k-1} - 1}{t-1}(t-1)\tilde{c}(t)$$
$$= \frac{(1 + (t-1)\tilde{c}(t))^k - 1}{t-1}.$$

(ii) The proof follows immediately from (i) bearing in mind that

$$T(UT_k(K)) = T^k(UT_1(K)) = T^k(K)$$

and $c_m(K) = 1$, i.e.,

$$\tilde{c}(K, t) = \sum_{m \geq 0} \frac{t^m}{m!} = e^t. \qquad \square$$

2.2 Background on Representation Theory of Groups

In this section we summarize briefly the necessary background on representation theory of groups. A systematic exposition of the material can be found, e.g., in [L] or [CR] for the general representation theory and in [JK, Mc] and [We2] for the representations of the symmetric and general linear groups. We use essentially that the base field K is of characteristic 0. (Some of the results are not true over fields of prime characteristic.) We assume that the elements of S_m act on the set $\{1, \ldots, m\}$ as functions and $(\sigma\tau)(i) = \sigma(\tau(i))$.

By the theorem of Maschke, the finite dimensional representations of any finite group G are completely reducible. Equivalently, the group algebra KG is semisimple and is isomorphic to the direct sum

$$M_{d_1}(D_1) \oplus \cdots \oplus M_{d_r}(D_r), \qquad (2.1)$$

where $M_{d_i}(D_i)$ are matrix algebras over the finite dimensional division algebras D_i, $i = 1, \ldots, r$. Every finite dimensional left G-module is a direct sum of irreducible (or simple) G-modules. Every irreducible G-module is isomorphic to a minimal left ideal of KG (and hence to a minimal left ideal of some $M_{d_i}(D_i)$), where G acts on KG by left multiplication.

In the special case of the symmetric group all division algebras in (2.1) coincide with K and

$$KS_m \cong M_{d_1}(K) \oplus \cdots \oplus M_{d_r}(K).$$

The nonisomorphic irreducible representations of the symmetric group (and hence the left irreducible S_m-modules) are in one-to-one correspondence with the conjugacy classes of S_m and are described in terms of partitions and Young diagrams.

Definition 2.2.1. A *partition* of the nonnegative integer m (notation $\lambda \vdash m$ or $|\lambda| = m$) is a sequence of integers $\lambda = (\lambda_1, \ldots, \lambda_r)$ such that

$$\lambda_1 \geq \cdots \geq \lambda_r \geq 0 \quad \text{and} \quad \lambda_1 + \cdots + \lambda_r = m.$$

We assume that two partitions $\lambda = (\lambda_1, \ldots, \lambda_r)$ and $\mu = (\mu_1, \ldots, \mu_s)$ are equal if

$$\lambda_1 = \mu_1, \ldots, \lambda_k = \mu_k, \lambda_{k+1} = \cdots = \lambda_r = \mu_{k+1} = \cdots = \mu_s = 0.$$

When $\lambda = (\lambda_1, \ldots, \lambda_{k_1 + \cdots + k_p})$ and

$$\lambda_1 = \cdots = \lambda_{k_1} = \mu_1, \ldots, \lambda_{k_1 + \cdots + k_{p-1} + 1} = \cdots = \lambda_{k_1 + \cdots + k_p} = \mu_p,$$

we accept the notation

$$\lambda = (\mu_1^{k_1}, \ldots, \mu_p^{k_p}).$$

There is a natural correspondence between the partitions of m and the conjugacy classes of S_m: If $\lambda = (\lambda_1, \lambda_2, \ldots, \lambda_r) \vdash m$, then the corresponding conjugacy class consists of all permutations σ of S_m which can be presented as products of independent cycles

$$\sigma = (i_1 \ldots i_{\lambda_1})(j_1 \ldots j_{\lambda_2}) \cdots (k_1 \ldots k_{\lambda_r}).$$

Definition 2.2.2. The *Young diagram* $[\lambda]$ of the partition $\lambda = (\lambda_1, \ldots, \lambda_r)$ is the set of all knots (points) $(i, j) \in \mathbb{Z}^2$, such that $1 \leq j \leq \lambda_i$, $i = 1, \ldots, r$.

It is convenient to present the Young diagrams graphically as follows. We replace the knots with square boxes such that the first coordinate i (the index of the row) increases from top to bottom and the second coordinate j (the index of the column) increases from left to right. For example, the diagram of the partition $\lambda = (5, 3^2, 2)$ is given in the figure below.

$$[\lambda] \quad = \quad$$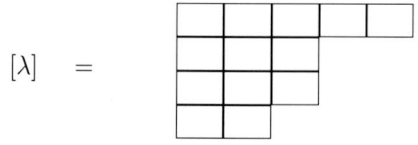

Definition 2.2.3. (i) A *Young tableau* T_λ of the diagram $[\lambda]$ with m boxes is a filling of the boxes of $[\lambda]$ with the positive integers $1, 2, \ldots, m$ without repetitions. If λ is a partition of m and $\tau \in S_m$, we denote by $T_\lambda(\tau)$ the tableau such that its first column contains the integers $\tau(1), \ldots, \tau(k_1)$ written in this order from top to bottom, the second column contains consequently written $\tau(k_1+1), \ldots, \tau(k_1+k_2)$, etc.

(ii) The tableau T_λ is called *standard*, if the integers written in each column and each row increase, respectively, from top to bottom and from left to right.

For example, for $\lambda = (4, 3, 1)$,

$$\tau = \begin{pmatrix} 1 & 2 & 3 & 4 & 5 & 6 & 7 & 8 \\ 3 & 4 & 5 & 8 & 2 & 1 & 6 & 7 \end{pmatrix}, \quad \tau_1 = \begin{pmatrix} 1 & 2 & 3 & 4 & 5 & 6 & 7 & 8 \\ 1 & 3 & 6 & 2 & 4 & 5 & 7 & 8 \end{pmatrix},$$

the tableau $T_\lambda(\tau)$ is not standard and the tableau $T_\lambda(\tau_1)$ is standard:

$$T_\lambda(\tau) = \begin{array}{|c|c|c|c|} \hline 3 & 8 & 1 & 7 \\ \hline 4 & 2 & 6 \\ \cline{1-3} 5 \\ \cline{1-1} \end{array} \qquad T_\lambda(\tau_1) = \begin{array}{|c|c|c|c|} \hline 1 & 2 & 5 & 8 \\ \hline 3 & 4 & 7 \\ \cline{1-3} 6 \\ \cline{1-1} \end{array}$$

Definition 2.2.4. Let $\lambda \vdash m$, $\tau \in S_m$ and let $T = T_\lambda(\tau)$ be the corresponding Young tableau. The *row stabilizer* of T is the subgroup $R(T)$ of all permutations ρ in S_m, such that i and $\rho(i)$ are in the same row of T, $i = 1, \ldots, m$. Similarly one defines the *column stabilizer* of T.

For the above example,

$$T = T_\lambda(\tau), \quad \lambda = (4, 3, 1), \quad \tau = \begin{pmatrix} 1 & 2 & 3 & 4 & 5 & 6 & 7 & 8 \\ 3 & 4 & 5 & 8 & 2 & 1 & 6 & 7 \end{pmatrix},$$

the row stabilizer $R(T)$ is the subgroup $S_4 \times S_3 \times S_1$ of S_8, where S_4, S_3 and S_1 act respectively on the sets $\{3, 8, 1, 7\}$, $\{4, 2, 6\}$ and $\{5\}$.

For each partition λ of m we denote by $M(\lambda)$ and χ_λ the corresponding irreducible S_m-module and its character, respectively.

Theorem 2.2.5. (i) *Let* $\lambda = (\lambda_1, \ldots, \lambda_r)$ *be a partition of* m, $\tau \in S_m$ *and let* $T = T_\lambda(\tau)$ *be the corresponding Young tableau. Up to a multiplicative constant the element of* KS_m

$$e_T = \sum_{\rho \in R(T)} \sum_{\gamma \in C(T)} (\text{sign } \gamma) \rho \gamma$$

is a minimal idempotent which generates a submodule of KS_m *isomorphic to* $M(\lambda)$.

(ii) *The sum of all left* S_m-*modules* $KS_m e_T$, *where* T *runs on the set of standard* λ-*tableaux, is direct. It is equal to the minimal two-sided ideal* $I(\lambda)$ *of* KS_m *corresponding to* λ, *and*

$$KS_m = \bigoplus_{\lambda \vdash m} I(\lambda).$$

(iii) *The dimension* $\dim M(\lambda)$ *of* $M(\lambda)$ *is given by the hook formula*

$$\dim M(\lambda) = \frac{m!}{\prod(\lambda_i + \lambda'_j - i - j + 1)},$$

where $\lambda'_1, \ldots, \lambda'_{r'}$ *are the lengths of the columns of* $[\lambda]$ *and the product in the denominator is on all boxes of* $[\lambda]$. *The dimension* $\dim M(\lambda)$ *is equal also to the number of standard* λ-*tableaux* $T_\lambda(\tau)$, $\tau \in S_m$.

For example, if $\lambda = (m)$, then the diagram of λ has one row only and for any (m)-tableau T

$$R(T) = S_m, \quad C(T) = 1.$$

Hence the one-dimensional trivial S_m-module $M(m)$ is spanned by the element

$$e_T = \sum_{\rho \in S_m} \rho.$$

In the other extreme case $\lambda = (1^m)$ we have $R(T) = 1$, $C(T) = S_m$. The one-dimensional S_m-module $M(1^m)$ corresponds to the sign representation and is spanned by

$$e_T = \sum_{\gamma \in S_m} (\text{sign } \gamma)\gamma.$$

If H is a subgroup of the finite group G, and if W and V are respectively G- and H-modules, then we denote by $W \downarrow H$ the module W considered as an H-module and by $V \uparrow G$ the G-module *induced* by V. Recall that $V \uparrow G = KG \otimes_{KH} V$. If one observes that $V \subseteq V \uparrow G$ as H-modules via the embedding $V \to KG \otimes_{KH} V$, $v \to 1 \otimes v$, then $V \uparrow G$ has the following universal property. For every G-module W' and for every homomorphism of H-modules $\varphi : V \longrightarrow W' \downarrow H$, there exists a unique homomorphism of G-modules $\psi : V \uparrow G \longrightarrow W'$ which extends φ.

Identifying S_{m-1} with the subgroup of S_m fixing the symbol m, the following *branching theorem* describes $M(\lambda) \downarrow S_{m-1}$, $\lambda \vdash m$, and $M(\mu) \uparrow S_m$, $\mu \vdash m - 1$. Parts (i) and (ii) are equivalent by the Frobenius reciprocity low.

Theorem 2.2.6. *Let* $\lambda \vdash m$, $\mu \vdash m - 1$. *Then:*

(i) $M(\lambda) \downarrow S_{m-1} \cong \bigoplus M(\mu^{(i)})$, *where the direct sum runs on all partitions* $\mu^{(i)}$ *of* $m - 1$ *such that their diagrams* $[\mu^{(i)}]$ *are obtained by deleting one box of the diagram* $[\lambda]$.

(ii) $M(\mu) \uparrow S_m \cong \bigoplus M(\lambda^{(j)})$, *where the summation is on all partitions* $\lambda^{(j)}$ *of* m *such that their diagrams* $[\lambda^{(j)}]$ *are obtained by adding one box to the diagram* $[\mu]$.

For example, if $m = 8$, $\mu = (3^2, 1)$, then

$$M(3^2, 1) \uparrow S_8 \cong M(4, 3, 1) \oplus M(3^2, 2) \oplus M(3^2, 1^2).$$

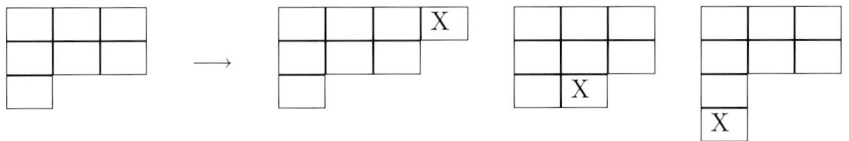

The branching theorem implies the following corollary.

Theorem 2.2.7. *Let $k < m$ and let μ be a partition of k. Then the two-sided ideal of KS_m generated by the ideal $I(\mu)$ of $KS_k \subset KS_m$ is a direct sum of all $I(\lambda)$, where $[\lambda]$ runs on the set of all diagrams with m boxes containing the diagram $[\mu]$.*

We shall need the following special case in Chapter 5 devoted to invariant theory of matrices.

Corollary 2.2.8. *If $k < m$, then the two-sided ideal of KS_m generated by*

$$e_T = \sum_{\sigma \in S_k} (\text{sign } \sigma)\sigma,$$

where T is the only standard μ-tableau for $\mu = (1^k)$, is equal to the sum of all $I(\lambda)$ where $\lambda = (\lambda_1, \ldots, \lambda_r) \vdash m$ and $\lambda_k \neq 0$.

Let W be a vector space. We denote by $GL(W)$ the *general linear group* of W, i.e., the group of invertible linear operators acting on W. When dim $W = d < \infty$, we write

$$GL_d = GL_d(K) = GL(W)$$

and, for a fixed basis $\{e_1, \ldots, e_d\}$ of W, we identify GL_d with the group of invertible $d \times d$ matrices with entries from K.

Definition 2.2.9. (i) A representation of the general linear group GL_d

$$\varphi : GL_d \longrightarrow GL_s = GL(W)$$

is called *polynomial* (and W is a *polynomial GL_d-module*), if the entries $\varphi_{pq}(g)$ of the $s \times s$ matrices $\varphi(g) \in GL_s$ are polynomial functions of the entries a_{ij} for all $d \times d$ matrices $g = (a_{ij}) \in GL_d$. When all φ_{pq} are homogeneous polynomials of degree m, then φ is a *homogeneous representation* of degree m.

(ii) Let

$$D_d = \{g \in GL_d \mid g = g(b_1, \ldots, b_d) = b_1 e_{11} + \cdots + b_d e_{dd}\}$$

be the subgroup of diagonal matrices of GL_d. For every d-tuple $\alpha = (\alpha_1, \ldots, \alpha_d)$ of integers, we define the *homogeneous component of weight α* (or *of degree α*) of the GL_d-module W by

$$W^\alpha = \{w \in W \mid g(b_1, \ldots, b_d)(w) = b_1^{\alpha_1} \cdots b_d^{\alpha_d} w \text{ for all } d \in D_d\}.$$

It is known that polynomial representations of general linear groups behave as representations of finite groups and have many common features with representations of symmetric groups.

Theorem 2.2.10. *Let $\varphi : GL_d \longrightarrow GL_s = GL(W)$ be a polynomial representation of GL_d. Then:*

(i) *The GL_d-module W is completely reducible and is a direct sum of homogeneous polynomial modules.*

(ii) *As a vector space W is a direct sum of its homogeneous components W^α, $\alpha \in \mathbb{Z}^d$.*

(iii) *The Hilbert series of W*

$$H(W) = H(W, t_1, \ldots, t_d) = \sum (\dim W^\alpha) t_1^{\alpha_1} \ldots t_d^{\alpha_d}$$

is a symmetric function of $t_1, \ldots, t_d$. If W is homogeneous of degree m, then $H(W)$ is also homogeneous of degree m.

(iv) *Two polynomial GL_d-modules W_1 and W_2 are isomorphic if and only if $H(W_1) = H(W_2)$. Besides,*

$$H(W_1 \oplus W_2) = H(W_1) + H(W_2), \quad H(W_1 \otimes W_2) = H(W_1) \cdot H(W_2).$$

Notice that by (iv) the Hilbert series of the polynomial GL_d-module W plays the role of the character of W.

The description of the irreducible polynomial representations of GL_d is the following.

Theorem 2.2.11. (i) *The irreducible polynomial representations of GL_d are in a one-to-one correspondence with the partitions $\lambda = (\lambda_1, \ldots, \lambda_d)$ in not more than d parts. We denote by $W_d(\lambda)$ the irreducible GL_d-module corresponding to λ.*

(ii) *The dimension of $W_d(\lambda)$ is given by the formula*

$$\dim W_d(\lambda_1, \ldots, \lambda_d) = \prod_{m \geq i > j \geq 1} \frac{\lambda_j - \lambda_i + i - j}{i - j}.$$

(iii) *Let*

$$V(\lambda_1, \ldots, \lambda_d) = \begin{vmatrix} t_1^{\lambda_1} & t_2^{\lambda_1} & \cdots & t_d^{\lambda_1} \\ t_1^{\lambda_2} & t_2^{\lambda_2} & \cdots & t_d^{\lambda_2} \\ \vdots & \vdots & \ddots & \vdots \\ t_1^{\lambda_d} & t_2^{\lambda_d} & \cdots & t_d^{\lambda_d} \end{vmatrix}.$$

Then the symmetric polynomial

$$S_\lambda(t_1, \ldots, t_d) = H(W_d(\lambda), t_1, \ldots, t_d) = \sum (\dim (W_d(\lambda))^\alpha) t_1^{\alpha_1} \cdots t_d^{\alpha_d},$$

called the Schur function of $W_d(\lambda)$ (and of λ if we assume that d is fixed), is expressed as

$$S_\lambda(t_1, \ldots, t_d) = \frac{V(\lambda_1 + d - 1, \lambda_2 + d - 2, \ldots, \lambda_{d-1} + 1, \lambda_d)}{V(d-1, d-2, \ldots, 1, 0)}.$$

(iv) Up to a multiplicative constant, there exists a unique nonzero element of weight λ in $W_d(\lambda)$. It is called the highest weight vector of $W_d(\lambda)$ and is characterized by the property that is invariant under the action of the subgroup $T_d(K)$ of GL_d of all upper triangular matrices with identity diagonal.

(v) Let W' and W'' be two submodules isomorphic to $W_d(\lambda)$ with highest weight vectors respectively w' and w''. If $\varphi : W' \longrightarrow W''$ is a GL_d-module isomorphism, then $\varphi(w') = \alpha w''$ for some nonzero $\alpha \in K$. Conversely, for every $0 \neq \alpha \in K$, there exists a unique GL_d-module isomorphism $\varphi : W' \longrightarrow W''$ such that $\varphi(w') = \alpha w''$.

Let V_d be a d-dimensional vector space with basis $\{x_1, \ldots, x_d\}$ and with the canonical action of GL_d, i.e., $GL_d = GL(V_d)$. The general linear group GL_d acts diagonally on the tensor algebra

$$T(V_d) = \bigoplus_{m \geq 0} V_d^{\otimes m},$$

i.e., for every $g \in GL_d$ and $x_{i_1} \otimes \cdots \otimes x_{i_m} \in V_d^{\otimes m}$

$$g(x_{i_1} \otimes \cdots \otimes x_{i_m}) = g(x_{i_1}) \otimes \cdots \otimes g(x_{i_m}).$$

It is easy to see that the multihomogeneous component of weight $(m_1, \ldots, m_d)$ of the GL_d-module $T(V_d)$ coincides with the vector subspace spanned by the tensors $x_{i_1} \otimes \cdots \otimes x_{i_m}$ of degree m_j with respect to x_j.

Theorem 2.2.12. Let $\lambda = (\lambda_1, \ldots, \lambda_d)$ be a partition of m.

(i) The irreducible polynomial GL_d-module $W_d(\lambda)$ is isomorphic to a submodule of the m-th tensor power $V_d^{\otimes m}$ of V_d and

$$V_d^{\otimes m} \cong \bigoplus (\dim M(\lambda)) W_d(\lambda), \qquad (2.2)$$

where $\dim M(\lambda)$ is the dimension of the corresponding S_m-module and the summation is on all partitions $\lambda = (\lambda_1, \ldots, \lambda_d) \vdash m$.

(ii) Let $d_\lambda = \dim M(\lambda)$ and let $W_1 \oplus \cdots \oplus W_{d_\lambda}$ be d_λ direct summands isomorphic to $W_d(\lambda)$ in decomposition (2.2) of $V_d^{\otimes m}$, and let $w_1, \ldots, w_{d_\lambda}$ be the corresponding highest weight vectors. Then the highest weight vector w of every $W_d(\lambda) \subset V_d^{\otimes m}$ is equal to a linear combination with nonzero coefficients:

$$w = a_1 w_1 + \cdots + a_{d_\lambda} w_{d_\lambda}, \quad a_i \in K.$$

For concrete computations with GL_d-submodules of $V_d^{\otimes m}$ one can use the explicit form of the highest weight vectors of the irreducible components in the decomposition (2.2). For details, see, e.g., [Dr10].

2.3   Group Representations and PI-algebras

In this section we shall describe some applications of representation theory of symmetric and general linear groups to the theory of PI-algebras. For more applications and for the possibilities for explicit computations see the book of the author [Dr10].

The following actions of the symmetric and the general linear groups are in the foundation of the methods and the results exposed in the present lectures.

The symmetric group S_m acts from the left on the set P_m of all multilinear homogeneous elements of degree m in $K\langle X \rangle$ by the rule

$$\sigma(x_{i_1} \cdots x_{i_m}) = x_{\sigma(i_1)} \cdots x_{\sigma(i_m)}, \quad x_{i_1} \cdots x_{i_m} \in P_m, \ \sigma \in S_m.$$

Then the left S_m-module P_m is isomorphic to the group algebra KS_m with the natural action of S_m by left multiplication.

The free algebra $A_d = K\langle x_1, \dots, x_d \rangle$ is isomorphic to the tensor algebra of the vector space V_d with basis $\{x_1, \dots, x_d\}$ and the action of GL_d is extended diagonally on A_d:

$$g(x_{i_1} \cdots x_{i_m}) = g(x_{i_1}) \cdots g(x_{i_m}), \quad x_{i_1} \cdots x_{i_m} \in A_d, g \in GL_d.$$

Then A_d is isomorphic to the tensor algebra of V_d also as a GL_d-module.

Although the application of representation theory of the symmetric groups to the study of PI-algebras started with the papers by Malcev [Ma] and Specht [Sp] in 1950, its systematic use for obtaining deep results with this method commenced in a series of papers by Regev in the 1970s, see [Rw1] for detailed bibliography of papers published in the 1970s.

Proposition 2.3.1. *For every PI-algebra R and for every $m \geq 0$, the set $P_m \cap T(R)$ is a submodule of the S_m-module P_m. Therefore,*

$$P_m(R) = P_m/(P_m \cap T(R))$$

is an S_m-module, a factor module of P_m.

Definition 2.3.2. Let R be a PI-algebra. The S_m-character

$$\chi_m(R) = \chi_{S_m}(P_m(R)), \quad m = 0, 1, 2, \dots,$$

is called the *m-th cocharacter* of R. Similarly one defines cocharacters for a variety of algebras $\mathfrak{U}$ using $P_m(\mathfrak{U}) = P_n/(P_n \cap T(\mathfrak{U}))$.

The representations of the symmetric group and the polynomial representations of the general linear group are equivalent, and they have been used *simultaneously* in many branches of mathematics. This happened incidentally also in the theory of PI-algebras. The systematic simultaneous application of these representations started with the papers by Berele [Be1] and the author [Dr1] in the early 1980s.

Proposition 2.3.3. *Let $\mathfrak{U}$ be a variety of algebras with T-ideal $U = T(\mathfrak{U})$ and let $d \geq 1$.*

(i) *The set $A_d \cap U$ is a submodule of the GL_d-module $A_d = K\langle x_1, \ldots, x_d \rangle$. Hence, the relatively free algebra $F_d(\mathfrak{U})$ is a factor module of A_d.*

(ii) *If*

$$F_d(\mathfrak{U}) \cong \bigoplus m'_\lambda W_d(\lambda_1, \ldots, \lambda_d)$$

is the decomposition of the GL_d-module $F_d(\mathfrak{U})$, then

$$H(F_d(\mathfrak{U}), t_1, \ldots, t_d) = \sum m'_\lambda S_\lambda(t_1, \ldots, t_d).$$

The multiplicities m'_λ, $\lambda = (\lambda_1, \ldots, \lambda_d)$, are uniquely determined by the Hilbert series $H(F_d(\mathfrak{U}), t_1, \ldots, t_d)$.

The following important theorem shows that for any PI-algebra R and the corresponding variety $\mathfrak{U} = \text{var } R$ the module structures of the S_m-modules $P_m(\mathfrak{U})$ and the GL_d-module $F_d(\mathfrak{U})$ coincide. It was proved by Berele [Be1], Theorem 2.7, and independently and with other methods by the author [Dr1], Remark 1.5 and [Dr3], Lemma 2.3. Later the proof of Berele was modified by Formanek [F3].

Theorem 2.3.4. *Let $\mathfrak{U}$ be a variety of algebras and let*

$$P_m(\mathfrak{U}) \cong \bigoplus_{\lambda \vdash m} m_\lambda M(\lambda), \quad m = 0, 1, 2, \ldots,$$

$$F_d(\mathfrak{U}) \cong \bigoplus_{m \geq 0} \bigoplus_{\lambda \vdash m} m'_\lambda W_d(\lambda).$$

Then $m'_\lambda = m_\lambda$ for every $\lambda = (\lambda_1, \ldots, \lambda_d)$ and $m'_\lambda = 0$ when $\lambda_{d+1} > 0$. Equivalently, if

$$\chi_m(\mathfrak{U}) = \sum_{\lambda \vdash m} m_\lambda \chi_\lambda,$$

then

$$H(F_d(\mathfrak{U}), t_1, \ldots, t_d) = \sum_{m \geq 0} \sum_{\lambda \vdash m} m_\lambda S_\lambda(t_1, \ldots, t_d),$$

where in the Hilbert series the summation is only on the partitions with $\lambda_{d+1} = 0$.

Example 2.3.5. By Lemma 1.4.1, the T-ideal $T(R)$ of any commutative unitary algebra R coincides with the commutator ideal of $K\langle X \rangle$. The algebra R generates the variety $\mathfrak{A}$ of all commutative algebras. Then $F_d(\mathfrak{A}) \cong K[x_1, \ldots, x_d]$. The multilinear component $P_m(\mathfrak{A})$ is one-dimensional and is spanned by $x_1 \cdots x_m$. Since

$$x_1 \cdots x_m = \frac{1}{m!} \left(\sum_{\rho \in S_m} \rho \right) x_1 \cdots x_m$$

in $F(\mathfrak{A})$, we obtain that $P_m(\mathfrak{A}) \cong M(m)$.

One can obtain the same presentation using the correspondence between the Hilbert series of relatively free algebras (their presentation as formal sums of Schur functions), the GL_d-module decomposition of relatively free algebras and the (multilinear) cocharacters of the corresponding T-ideals. It is easy to see that

$$H(F_d(\mathfrak{A}), t_1, \ldots, t_d) = H(K[x_1, \ldots, x_d], t_1, \ldots, t_d)$$

$$= \prod_{i=1}^{d} \frac{1}{1 - t_i} = \sum_{m \geq 0} S_{(m)}(t_1, \ldots, t_d).$$

Hence

$$F_d(\mathfrak{A}) = \bigoplus_{m \geq 0} W_d(m)$$

for all $d \geq 1$ and

$$\chi_m(\mathfrak{A}) = \chi_{(m)}.$$

Example 2.3.6. The theorem of Krakowski and Regev [KR] gives the following decomposition of the S_m-cocharacter of the Grassmann algebra:

$$\chi_m(E) = \sum_{k=1}^{m} \chi_{(k,1^{m-k})}, \quad m \geq 1.$$

An easy description of all the numerical invariants of the polynomial identities of the Grassmann algebra (also over a finite dimensional vector space) based on the so called *proper* polynomial identities, is given for example in [DV2] or [Dr10].

In Chapter 3 dealing with polynomial identities of matrices and in other places in the present lecture notes, we shall survey concrete results on cocharacters of matrix algebras and other related objects. We want to mention that Theorem 2.3.4 works also for the extreme case of the variety $\mathfrak{O}$ consisting of all associative algebras. In this case $F(\mathfrak{O}) = K\langle X \rangle$,

$$\chi_m(\mathfrak{O}) = \sum_{\lambda \vdash m} d_\lambda \chi_\lambda, \quad d_\lambda = \dim M(\lambda),$$

$$F_d(\mathfrak{O}) = \bigoplus_{m \geq 0} \bigoplus_{\lambda \vdash m} d_\lambda W_d(\lambda),$$

$$H(F_d(\mathfrak{O}), t_1, \ldots, t_d) = \frac{1}{1 - (t_1 + \cdots + t_d)} = \sum_{m \geq 0} \sum_{\lambda \vdash m} d_\lambda S_\lambda(t_1, \ldots, t_d).$$

We shall complete this section with two theorems given without proofs. They provide some information on the shape of the diagrams for which the related irreducible S_m-characters participate in the cocharacter sequence with nonzero multiplicities.

Theorem 2.3.7. (Regev [Re3]) *The algebra R satisfies the Capelli identity in n skew-symmetric variables*

$$d_n(x_1, \ldots, x_n; y_1, \ldots, y_{n-1}) = \sum_{\sigma \in S_n} (\text{sign } \sigma) x_{\sigma(1)} y_1 x_{\sigma(2)} y_2 \cdots y_{n-1} x_{\sigma(n)} = 0$$

if and only if its cocharacter sequence is decomposed as

$$\chi_m(R) = \sum_{\lambda_n = 0} m_\lambda \chi_\lambda, \quad m = 0, 1, 2, \ldots,$$

i.e., the nonzero multiplicities correspond to partitions in less than n parts.

The proof follows easily from the concrete form of the idempotents in the group algebra KS_m. If $\lambda = (\lambda_1, \ldots, \lambda_k)$, $k > n$, and $\lambda_k \neq 0$, then, for each λ-tableau $T = T_\lambda(\tau)$ the element e_T is a linear combination (with coefficients from KS_m) of skew-symmetric sums on a symmetric group of degree k. The same holds for the element $e_T(x_1 \cdots x_m)$ and this means that the identity $e_T(x_1 \cdots x_m) = 0$ is a consequence of the Capelli identity. Hence all S_m-modules $M(\lambda)$ with $\lambda = (\lambda_1, \ldots, \lambda_k)$, $k > n$ and $\lambda_k \neq 0$, belong to the T-ideal generated by the Capelli identity and the multiplicity m_λ in the cocharacter sequence is equal to 0.

Corollary 2.3.8. *If* $\dim R < n$, *then*

$$\chi_m(R) = \sum_{\lambda_n = 0} m_\lambda \chi_\lambda, \quad \lambda \vdash m, \ m = 0, 1, 2, \ldots.$$

The following theorem of Amitsur and Regev [AR] is much deeper and uses essentially the theorem of Regev for the exponential growth of the codimension sequence of a PI-algebra.

Theorem 2.3.9. (Amitsur and Regev [AR]) *For every PI-algebra R the Young diagrams corresponding to the irreducible characters participating in the cocharacter sequence of R are in a hook, i.e., there exist integers k, l such that*

$$\chi_m(R) = \sum m_\lambda \chi_\lambda, \quad m = 0, 1, 2, \ldots,$$

and the summation is on all $\lambda = (\lambda_1, \ldots, \lambda_p) \vdash m$ *satisfying* $\lambda_{k+1} \leq l$.

Chapter 3

The Amitsur–Levitzki Theorem

3.1 The Amitsur–Levitzki Theorem and Related Results

Since the matrix algebras are among the most important objects in algebra and its applications, their polynomial identities have been an attractive object for study from the very origins of the theory of PI-algebras.

One of the first results is the famous *Amitsur–Levitzki theorem* which states that the $n \times n$ matrix algebra satisfies the standard identity of degree $2n$,

$$s_{2n}(x_1, \ldots, x_{2n}) = \sum_{\sigma \in S_{2n}} (\text{sign } \sigma) x_{\sigma(1)} \cdots x_{\sigma(2n)} = 0.$$

Up to a multiplicative constant, this is the only multilinear polynomial identity of degree $\leq 2n$ for $M_n(K)$. There are five essentially different proofs of the Amitsur–Levitzki theorem. The original proof [AL] is based on inductive combinatorial arguments; with some technical improvements it can be found, for example, in the book by Passmann [Pa], p. 175. There is a graph theoretical proof of Swan [Sw] which treats the nonzero products of matrix units as trails of *Eulerian graphs* (oriented graphs such that there exists a trail passing through all edges exactly once). Recently, the basic idea of this proof was used by Szigeti, Tuza and Révész [STR] to give another proof and to show that the matrix algebras satisfy also other interesting polynomial identities. Kostant [Ks] gave a cohomological proof. Finally, there are two other proofs of the Amitsur–Levitzki theorem due to Razmyslov [Ra3] and Rosset [Ro]. Here we shall give the proof of Razmyslov. The next lemma is well known.

Lemma 3.1.1. (i) *The $n \times n$ matrix algebra $M_n(K)$ does not satisfy a polynomial identity of degree less than $2n$.*

(ii) *If $f(x_1, \ldots, x_{2n}) = 0$ is a multilinear polynomial identity of degree $2n$ for $M_n(K)$, then*

$$f(x_1, \ldots, x_{2n}) = \alpha s_{2n}(x_1, \ldots, x_{2n}), \quad \alpha \in K.$$

Proof. (i) If $M_n(K)$ satisfies a polynomial identity of degree $k < 2n$, then it satisfies also a multilinear identity

$$f(x_1, \ldots, x_k) = \sum_{\sigma \in S_k} \alpha_\sigma x_{\sigma(1)} \cdots x_{\sigma(k)} = 0, \quad \alpha_\sigma \in K,$$

of degree k. We use *staircase arguments*:

$$f(e_{11}, e_{12}, e_{22}, e_{23}, \ldots, e_{pq}) = \alpha_\varepsilon e_{1p},$$

where $p = q$ or $p = q - 1$ depending on the parity of k, and ε is the identity permutation. Hence $\alpha_\varepsilon = 0$. It is easy to modify these arguments to show also that $\alpha_\sigma = 0$, for all permutations σ in S_k.

(ii) Let

$$f(x_1, \ldots, x_{2n}) = \sum_{\sigma \in S_{2n}} \alpha_\sigma x_{\sigma(1)} \cdots x_{\sigma(2n)} = 0$$

be a multilinear identity of degree $2n$ for $M_n(K)$. As in (i), we replace $x_1, \ldots, x_{2n}$ by matrix units. For example,

$$f(e_{11}, e_{11}, e_{12}, e_{22}, \ldots, e_{n-1,n}, e_{nn}) = (\alpha_\varepsilon + \alpha_{(12)})e_{1n} = 0$$

and, similarly, $\alpha_\sigma + \alpha_\tau = 0$ if τ is obtained from σ by transposition of two consequent elements $\tau(p)$ and $\tau(p+1)$. This implies $\alpha_\sigma = (\text{sign } \sigma)\alpha_\varepsilon$ for all $\sigma \in S_{2n}$. $\square$

Lemma 3.1.2. *If the $n \times n$ matrix algebra $M_n(\mathbb{Q})$ with entries from the field $\mathbb{Q}$ of all rational numbers satisfies the standard identity of degree $2n$, then the algebra $M_n(K)$ over any field K also satisfies $s_{2n}(x_1, \ldots, x_{2n}) = 0$.*

Proof. Let

$$r_p = \sum_{i,j=1}^{n} \alpha_{ij}^{(p)} e_{ij}, \quad \alpha_{ij}^{(p)} \in K, \quad p = 1, \ldots, 2n,$$

be matrices in $M_n(K)$. Since the standard polynomial $s_{2n}(x_1, \ldots, x_{2n})$ is multilinear and skew-symmetric, we obtain that $s_{2n}(r_1, \ldots, r_{2n})$ is a linear combination of $s_{2n}(e_{i_1 j_1}, \ldots, e_{i_{2n} j_{2n}})$ and is equal to 0 because we assumed that $s_{2n}(x_1, \ldots, x_{2n}) = 0$ for $M_n(\mathbb{Z}) \subset M_n(\mathbb{Q})$. $\square$

Lemma 3.1.3. (i) *Let $\xi_i, \ldots, \xi_n$ be the eigenvalues of the matrix $a \in M_n(K)$ and let $e_q = e_q(\xi_1, \ldots, \xi_n)$ be the q-th elementary symmetric polynomial in $\xi_1, \ldots, \xi_n$. Then*

$$a^n + \sum_{q=1}^{n} (-1)^q e_q(\xi_1, \ldots, \xi_n) a^{n-q} = 0,$$

$$\text{tr}(a^q) = \xi_1^q + \cdots + \xi_n^q.$$

(ii) *When K is of characteristic zero, the coefficients $e_q(\xi_1, \ldots, \xi_n)$ can be expressed as polynomials with rational coefficients of $\text{tr}(a^q)$, $q = 1, 2, \ldots, n$.*

Proof. (i) For the proof it is sufficient to apply the Cayley–Hamilton theorem to the matrix a written in its Jordan normal form.

(ii) Let

$$p_q = p_q(\xi_1, \ldots, \xi_n) = \xi_1^q + \cdots + \xi_n^q.$$

Applying the Newton formulas for $q \leq n$,

$$p_q - e_1 p_{q-1} + e_2 p_{q-2} + \cdots + (-1)^{q-1} e_{q-1} p_1 + (-1)^q q e_q = 0,$$

we can express $e_q(\xi_1, \ldots, \xi_n)$ as polynomials of $p_q(\xi_1, \ldots, \xi_n)$. $\qquad\square$

For example, for $n = 2$ we have

$$e_2(\xi_1, \xi_2) = \xi_1 \xi_2 = \frac{1}{2}((\xi_1 + \xi_2)^2 - (\xi_1^2 + \xi_2^2)) = \frac{1}{2}(p_1^2(\xi_1, \xi_2) - p_2(\xi_1, \xi_2)).$$

For any 2×2 matrix a with eigenvalues ξ_1, ξ_2, the Cayley–Hamilton theorem gives that

$$a^2 - e_1(\xi_1, \xi_2)a + e_2(\xi_1, \xi_2)e = 0,$$

$$a^2 - p_1(\xi_1, \xi_2)a + \frac{1}{2}(p_1^2(\xi_1, \xi_2) - p_2(\xi_1, \xi_2))e = 0,$$

$$a^2 - \mathrm{tr}(a)a + \frac{1}{2}(\mathrm{tr}^2(a) - \mathrm{tr}(a^2))e = 0.$$

Hence

$$x^2 - \mathrm{tr}(x)x + \frac{1}{2}(\mathrm{tr}^2(x) - \mathrm{tr}(x^2))e = 0$$

is a *trace polynomial identity* for $M_2(K)$.

Theorem 3.1.4. (Amitsur and Levitzki [AL]) *Over any field K, the $n \times n$ matrix algebra $M_n(K)$ satisfies the standard identity of degree $2n$. It does not satisfy polynomial identities of lower degree and, up to a multiplicative constant, $s_{2n} = 0$ is the only multilinear polynomial identity of degree $2n$ for $M_n(K)$.*

Proof. Below we give the proof of Razmyslov [Ra3]. For simplicity of the exposition we consider the case of 2×2 matrices only. The general case is similar and the only additional difficulties are of technical nature. By Lemma 3.1.2 we assume that $K = \mathbb{Q}$. By Lemma 3.1.3 the algebra $M_2(\mathbb{Q})$ satisfies the trace identity

$$x^2 - \mathrm{tr}(x)x + \frac{1}{2}(\mathrm{tr}^2(x) - \mathrm{tr}(x^2))e = 0.$$

We linearize this identity (it is easy to see that we can linearize not only ordinary but also trace polynomial identities) and obtain the identity

$$(y_1 y_2 + y_2 y_1) - (\mathrm{tr}(y_1)y_2 + \mathrm{tr}(y_2)y_1)$$
$$+ \frac{1}{2}((\mathrm{tr}(y_1)\mathrm{tr}(y_2) + \mathrm{tr}(y_2)\mathrm{tr}(y_1)) - \mathrm{tr}(y_1 y_2 + y_2 y_1))e = 0.$$

Since

$$\mathrm{tr}(y_1)\mathrm{tr}(y_2) = \mathrm{tr}(y_2)\mathrm{tr}(y_1), \ \mathrm{tr}(y_1 y_2) = \mathrm{tr}(y_2 y_1),$$

we observe that $M_2(\mathbb{Q})$ satisfies the multilinear trace identity

$$f(y_1, y_2) = (y_1 y_2 + y_2 y_1) - (\mathrm{tr}(y_1)y_2 + \mathrm{tr}(y_2)y_1) + (\mathrm{tr}(y_1)\mathrm{tr}(y_2) - \mathrm{tr}(y_1 y_2))e = 0.$$

Now we replace y_1 and y_2 respectively by $x_{\sigma(1)}x_{\sigma(2)}$ and $x_{\sigma(3)}x_{\sigma(4)}$ and take the alternating sum on $\sigma \in S_4$:

$$
\begin{aligned}
0 &= \sum_{\sigma \in S_4} (\text{sign } \sigma) f(x_{\sigma(1)}x_{\sigma(2)}, x_{\sigma(3)}x_{\sigma(4)}) \\
&= 2 \sum_{\sigma \in S_4} (\text{sign } \sigma)(x_{\sigma(1)}x_{\sigma(2)}x_{\sigma(3)}x_{\sigma(4)} - \text{tr}(x_{\sigma(1)}x_{\sigma(2)})x_{\sigma(3)}x_{\sigma(4)}) \\
&\quad + \sum_{\sigma \in S_4} (\text{sign } \sigma)(\text{tr}(x_{\sigma(1)}x_{\sigma(2)})\text{tr}(x_{\sigma(3)}x_{\sigma(4)}) - \text{tr}(x_{\sigma(1)}x_{\sigma(2)}x_{\sigma(3)}x_{\sigma(4)}))e.
\end{aligned}
$$

The trace is invariant under cyclic permutations, hence

$$\text{tr}(x_{\sigma(1)}x_{\sigma(2)}) = \text{tr}(x_{\sigma(2)}x_{\sigma(1)}),$$

$$\text{tr}(x_{\sigma(1)}x_{\sigma(2)}x_{\sigma(3)}x_{\sigma(4)}) = \text{tr}(x_{\sigma(2)}x_{\sigma(3)}x_{\sigma(4)}x_{\sigma(1)}).$$

On the other hand, in each of the pairs

$$(\sigma(1), \sigma(2), \sigma(3), \sigma(4)) \quad \text{and} \quad (\sigma(2), \sigma(1), \sigma(3), \sigma(4)),$$

$$(\sigma(1), \sigma(2), \sigma(3), \sigma(4)) \quad \text{and} \quad (\sigma(2), \sigma(3), \sigma(4), \sigma(1)),$$

the permutations are of different parity and the summands containing traces vanish in

$$\sum_{\sigma \in S_4} (\text{sign } \sigma) f(x_{\sigma(1)}x_{\sigma(2)}, x_{\sigma(3)}x_{\sigma(4)}) = 0.$$

Therefore, we obtain that

$$2s_4 = 2 \sum_{\sigma \in S_4} (\text{sign } \sigma)x_{\sigma(1)}x_{\sigma(2)}x_{\sigma(3)}x_{\sigma(4)} = 0,$$

and this completes the proof. $\qquad\square$

The following theorem contains a result obtained independently by Chang [Ch] and Giambruno and Sehgal [GS]. It answers a problem risen by Formanek.

Theorem 3.1.5. *The integer $2n$ is the minimal positive integer such that the matrix algebra $M_n(K)$ (char $K = 0$) satisfies the double Capelli identity*

$$f_{2n}(x_1, \ldots, x_{2n}; y_1, \ldots, y_{2n}) = \sum_{\sigma, \tau \in S_{2n}} (\text{sign } (\sigma\tau))x_{\sigma(1)}y_{\tau(1)} \cdots x_{\sigma(2n)}y_{\tau(2n)}.$$

Proof. Considering the *"double staircase"*

$$\bar{x}_1 = e_{11}, \bar{y}_1 = e_{12}, \bar{x}_2 = e_{22}, \bar{y}_2 = e_{23}, \ldots, \bar{x}_n = e_{nn},$$

$$\bar{y}_n = e_{nn}, \bar{x}_{n+1} = e_{n,n-1}, \bar{y}_{n+1} = e_{n-1,n-1}, \ldots, \bar{x}_{2n-1} = e_{21}, \bar{y}_{2n-1} = e_{11},$$

we observe that $f_{2n-1}(\bar{x}_1, \ldots, \bar{x}_{2n-1}; \bar{y}_1, \ldots, \bar{y}_{2n-1}) \neq 0$.

The following easy proof that $f_{2n} = 0$ is an identity for $M_n(K)$ is due to Domokos [Do1]. Let $u_1, \ldots, u_{2n}, v_1, \ldots, v_{2n}$ be $n \times n$ matrices. We form the $2n \times 2n$ block matrices

$$w_{2q-1} = \begin{pmatrix} 0 & u_q \\ 0 & 0 \end{pmatrix}, \ w_{2q} = \begin{pmatrix} 0 & 0 \\ v_q & 0 \end{pmatrix}, \ q = 1, \ldots, 2n.$$

Since the product $w_i w_j$ is not zero only if i and j are of different parity, we obtain that $s_{4n}(w_1, \ldots, w_{4n})$ is equal to the matrix

$$\begin{pmatrix} f_{2n}(u_1, \ldots, u_{2n}; v_1, \ldots, v_{2n}) & 0 \\ 0 & f_{2n}(v_1, \ldots, v_{2n}; u_1, \ldots, u_{2n}) \end{pmatrix}$$

and this is 0 in virtue of the Amitsur–Levitzki theorem for $M_{2n}(K)$. $\qquad \square$

The proof of Chang [Ch] is combinatorial and contains the stronger result that the double Capelli identity in Theorem 3.1.5 is a consequence of the standard identity $s_{2n} = 0$. The proof of Giambruno and Sehgal [GS] is based on the Rosset proof [Ro] of the Amitsur–Levitzki theorem. Another proof can be found in the paper by Szigeti, Tuza and Révész [STR].

By Example 1.1.5 (iv), the Capelli identity in $n^2 + 1$ skew-symmetric variables holds for the algebra $M_n(K)$. Now, we shall show that this is the Capelli identity of minimal degree satisfied by $M_n(K)$. We shall use this fact in the construction of central polynomials.

Lemma 3.1.6. *The algebra $M_n(K)$ does not satisfy the Capelli identity*

$$d_{n^2}(x_1, \ldots, x_{n^2}; y_1, \ldots, y_{n^2-1}) = 0.$$

Proof. We order the matrix units e_{ij} in an arbitrary way

$$e_{p_1 q_1}, e_{p_2 q_2}, \ldots, e_{p_{n^2} q_{n^2}}.$$

Direct calculations show that

$$e_{1p_1} d_{n^2}(e_{p_1 q_1}, e_{p_2 q_2}, \ldots, e_{p_{n^2} q_{n^2}}; e_{q_1 p_2}, \ldots, e_{q_{n^2-1} p_{n^2}}) e_{q_{n^2} 1} = e_{11} \neq 0. \qquad \square$$

Finally, we shall give another identity for $M_n(K)$ which does not follow from the standard identity.

Lemma 3.1.7. *The algebra $M_n(K)$ satisfies the identity of algebraicity*

$$a_n(x; y_1, \ldots, y_n) = d_{n+1}(1, x, x^2, \ldots, x^n; y_1, \ldots, y_n)$$
$$= \sum_{\sigma \in S_{n+1}} (\text{sign } \sigma) x^{\sigma(0)} y_1 x^{\sigma(1)} y_2 \cdots y_n x^{\sigma(n)} = 0,$$

where the symmetric group S_{n+1} acts on $\{0, 1, \ldots, n\}$, and the identity

$$s_n([x, y], [x^2, y], \ldots, [x^n, y]) = 0.$$

For $n \geq 2$ these identities do not follow from the standard identity $s_{2n} = 0$.

Proof. For any $n \times n$ matrix a with characteristic polynomial $f_a(\lambda)$, the Cayley–Hamilton theorem gives that

$$f_a(a) = a^n + \alpha_1 a^{n-1} + \cdots + \alpha_{n-1} a + \alpha_n e = 0,$$

where $\alpha_1, \ldots, \alpha_{n-1}, \alpha_n$ are constants in K. Hence the matrices $e, a, a^2, \ldots, a^n$ are linearly dependent and the Capelli identity

$$d_{n+1}(e, a, a^2, \ldots, a^n; b_1, \ldots, b_n) = 0$$

holds for all $b_1, \ldots, b_n \in M_n(K)$. Hence $M_n(K)$ satisfies the identity of algebraicity. Similarly, for any $b \in M_n(K)$,

$$0 = [f_a(a), b] = [a^n, b] + \alpha_1 [a^{n-1}, b] + \cdots + \alpha_{n-1}[a, b]$$

and we conclude that $[a, b], [a^2, b], \ldots, [a^n, b]$ are also linearly dependent. Hence $s_n([a, b], [a^2, b], \ldots, [a^n, b]) = 0$ and $M_n(K)$ satisfies the polynomial identity

$$s_n([x, y], [x^2, y], \ldots, [x^n, y]) = 0.$$

Further, we shall pay attention to the fact that for $k = 2p$ even, the standard polynomial can be rewritten as

$$s_{2p} = \frac{1}{2^p} \sum_{\sigma \in S_{2p}} (\text{sign } \sigma)[x_{\sigma(1)}, x_{\sigma(2)}] \cdots [x_{\sigma(2p-1)}, x_{\sigma(2p)}]$$

and vanishes if we replace one of the variables by 1.

Now, let us assume that the identity of algebraicity $a_n = 0$ follows from the standard identity $s_{2n} = 0$. Then the consequence $a_n(x; y, \ldots, y) = 0$ of $a_n(x; y_1, \ldots, y_n) = 0$ (obtained by the substitution $y_1 = \cdots = y_n = y$) also follows from $s_{2n} = 0$ and can be presented in the form

$$a_n(x, y, \ldots, y) = \sum \alpha_i u_i s_{2n}(v_{i_1}, \ldots, v_{i_{2n}}) w_i, \quad \alpha_i \in K,$$

where u_i, v_{i_j}, w_i are monomials in $K\langle x, y \rangle$. We may assume that the total degree of u_i, v_{i_j}, w_i in y is n and in x is $\frac{1}{2}n(n+1)$. This means that in the standard polynomial $s_{2n}(v_{i_1}, \ldots, v_{i_{2n}})$ not more than n monomials v_{i_j} contain y and the others are positive powers of x. Since the standard identity is skew-symmetric, we obtain that all monomials v_{i_j} are different and this gives that the total degree of x is not less than $1 + 2 + \cdots + n$. Hence, all monomials v_j containing y should be equal to y and, since $n \geq 2$, this is impossible. The arguments for the identity $s_n([x, y], [x^2, y], \ldots, [x^n, y]) = 0$ are similar. $\square$

3.2   Polynomial Identities of Low Degree — a Survey

In this section we shall survey some results and some well known problems and conjectures on concrete polynomial identities satisfied by the matrix algebra $M_n(K)$

over a field of characteristic 0. For more details, see the survey [F2] and the book [F6] by Formanek. See also the survey article by Benanti, Demmel, Drensky and Koev [BDDK] for the computational approach to the polynomial identities of $M_n(K)$. (Some of the facts in the theory have been obtained, or at least conjectured, as a result of heavy calculations by hand or computer.) Of course, one of the main problems in the theory of polynomial identities of matrices is the following.

Problem 3.2.1. *Find a basis of the polynomial identities for the algebra of $n \times n$ matrices, $n \geq 2$, over a field of characteristic 0.*

We have already mentioned that the complete answer is known for 2×2 matrices only, see [Ra1] and [Dr2]. Up to the present date, no explicit bases of the identities of $M_n(K)$, $n > 2$, are known.

In 1973, Leron [Le] proved that for $n > 2$ all polynomial identities of degree $2n + 1$ for $M_n(K)$ are consequences of the standard identity $s_{2n} = 0$. For $n = 3$ this result was improved by Drensky and Kasparian [DK1]. In 1983, they found all identities of degree 8 and showed that they are consequences of the standard identity $s_6 = 0$. These calculations (done by hand) were checked by computer by Bondari [Bo] in 1997 and by Vishne [Vi] in 2002.

The polynomial identity for 2×2 matrices

$$[[x, y]^2, x] = 0$$

is equal to the identity of algebraicity

$$a_2(x; y, y) = \sum_{\sigma \in S_3} (\text{sign } \sigma) x^{\sigma(0)} y x^{\sigma(1)} y x^{\sigma(2)} = 0.$$

Hence, by [Ra1] and [Dr2], the standard identity $s_4(x_1, x_2, x_3, x_4) = 0$ and the identity of algebraicity $a_2(x; y_1, y_2) = 0$ form a basis of the identities for $M_2(K)$.

There was a hope that a similar result holds for the matrices of any size and $s_{2n} = 0$ and $a_n(x; y_1, \ldots, y_n) = 0$ form a basis for $M_n(K)$ also for $n > 2$.

But for $n = 3$ several new identities of degree 9 were found by Okhitin [Ok1] and Domokos [Do2]. For $n > 3$ no new identities of degree $< \deg a_n = n(n+3)/2$ are known.

Problem 3.2.2. *Let $n \geq 4$.*

(i) *What is the minimal degree of the polynomial identities for $M_n(K)$ which do not follow from $s_{2n} = 0$? Is this degree equal to the degree of the identity of algebraicity $\deg a_n = n(n+3)/2$? (For $n = 2, 3$ this degree is equal to $\deg a_n$.)*

(ii) *What is the minimal degree of the polynomial identities in two variables for $M_n(K)$ which do not follow from $s_{2n} = 0$?*

(iii) *What is the minimal degree of the polynomial identities in two variables for $M_n(K)$?*

Remark 3.2.3. For n sufficiently large, $M_n(K)$ has polynomial identities in two variables which are consequences of the standard identity and are of lower degree than the identity of algebraicity (with $y_1 = \ldots = y_n = y$). For the proof, we order all monomials in two variables in $K\langle x, y \rangle$ by ascending degree, e.g.,

$$u_1 = x,\; u_2 = y,\; u_3 = x^2,\; u_4 = xy,\; u_5 = yx,\; u_6 = y^2, \ldots,$$

and, using these elements "economically", we construct a nonzero polynomial in $K\langle x, y \rangle$ of the form

$$s_{2n}(u_{i_1}, \ldots, u_{i_{2n}}), \quad i_1 < \cdots < i_{2n}.$$

For n sufficiently large, the degree of the obtained identity is less than the degree of the identity of algebraicity $\frac{1}{2}n(n + 3)$.

Based on the result of Leron [Le] and the tiny "experimental material" for $n = 3$, see [DK1] and also [Bo], [Vi], one may conjecture the following [BDDK]:

Conjecture 3.2.4. *For $n > 2$ all polynomial identities of degree $2n + 2$ for the matrix algebra $M_n(K)$ are consequences of the standard identity $s_{2n} = 0$.*

Benanti, Demmel, Drensky and Koev, see [BDDK], combined theoretical conclusions and computer experiments to confirm Conjecture 3.2.4 for $n = 4, 5$. They performed the computer calculations for the polynomial identities using 64 processors on the Cray T3E at the Lawrence Berkeley Lab. It took a total of about 8 hours to complete the computations for the identities of degree 12 for 5×5 matrices from start to finish. (In order to compare the results with the known facts, six hours of this time were spent for calculations which follow from other arguments.)

It is not clear how realistic is to hope that all polynomial identities of degree $< n(n + 3)/2$ for $M_n(K)$ follow from $s_{2n} = 0$. Nevertheless it seems reasonable to believe that the identities for $M_n(K)$ of degree $2n + k$ are consequences of $s_{2n} = 0$ for small k. Very probably, k increases at least linearly with the size n of the matrices.

We want to mention two other kinds of known identities. The first kind, described by Bergman [Bg2], is of the form $f(x, y_1, \ldots, y_n) = 0$ where f is linear in each variable $y_1, \ldots, y_n$. It turns out that all these identities are consequences of the identity of algebraicity. The second kind of identities, studied by Benediktovich and Zalesskij [BZ], consists of the identities of degree m for $M_n(K)$ which are skew-symmetric in $m - 1$ variables. All these identities are consequences of the standard identity $s_{2n} = 0$. Finally, there are also polynomial identities which are obtained from central polynomials and this will be the main topic of the next chapter.

We shall give some idea how one can (in principle) perform concrete calculations with polynomial identities. First we consider the problem how to find all polynomial identities of degree m for $M_n(K)$. The simplest (but not the best) algorithm uses that the polynomial identities of any algebra are equivalent to the multilinear identities and is the following. We consider the multilinear identity

$$f(x_1, \ldots, x_m) = \sum_{\sigma \in S_m} \xi_\sigma x_{\sigma(1)} \cdots x_{\sigma(m)} = 0 \tag{3.1}$$

with unknown coefficients ξ_σ, $\sigma \in S_m$. Since $f(x_1, \ldots, x_m) = 0$ is a polynomial identity for $M_n(K)$ if and only if $f(e_{i_1 j_1}, \ldots, e_{i_m j_m}) = 0$ for all matrix units $e_{i_p j_p}$, $p = 1, \ldots, m$, we consider each of the n^2 entries of

$$f(e_{i_1 j_1}, \ldots, e_{i_k j_m}) = \sum_{\sigma \in S_m} \xi_\sigma e_{i_{\sigma(1)} j_{\sigma(1)}} \cdots e_{i_{\sigma(m)} j_{\sigma(m)}} = 0 \qquad (3.2)$$

as a linear homogeneous equation. In this way, we obtain a system with $m!$ unknowns ξ_σ, $\sigma \in S_m$, and $n^2 \times (n^2)^m$ equations. The set of the solutions coincides with the set of multilinear identities of degree m.

If we want to show that all identities of degree m are consequences of a given system of multilinear polynomial identities $F = \{f_i(x_1, \ldots, x_{m_i}) = 0 \mid i \in I\}$, we may proceed in the following way. The vector space of the multilinear consequences of degree m of the system F is spanned by all polynomial identities of the form

$$u_0 f_i(u_1, \ldots, u_{m_i}) u_{m_i+1} = 0, \qquad (3.3)$$

where $u_0, u_1, \ldots, u_{m_i}, u_{m_i+1}$ are monomials of total degree m which involve exactly once each variable $x_1, \ldots, x_m$. Since each $f_i(x_1, \ldots, x_{m_i}) = 0$ is an identity for $M_n(K)$, the identities (3.3) presented in the form (3.1), are obtained as solutions of the system (3.2). Hence, all polynomial identities of degree m follow from the identities $f_i = 0$, $i \in I$, if and only if the dimension d_F of the vector space spanned by (3.3) coincides with the dimension d_S of the vector space of the solutions of (3.2). Since in the general case $d_F \leq d_S$, the algebra $M_n(K)$ satisfies identities of degree m which do not follow from the system F if and only if $d_F < d_S$. Then the difference $d_S - d_F$ gives the dimension of the "new identities" of degree m.

Instead to search for solutions of the whole system (3.2), we may consider only a sufficiently big part of the equations and to calculate that the dimension of the corresponding vector space of solutions is d'_S. Clearly $d'_S \geq d_S$. We also may consider a part of the consequences (3.3) and show that they generate a vector subspace of dimension d'_F (and $d'_F \leq d_F$). If, by chance, $d'_F = d'_S$, this implies that $d_F = d_S$ and all identities of degree m are consequences of the system F.

A similar strategy was applied by Leron [Le] to show that all polynomial identities of degree $2n + 1$ for the matrix algebra $M_n(K)$, $n > 2$, follow from the standard identity $s_{2n} = 0$. It turns out that both dimensions, of the consequences of the standard identity, and of all identities of degree $2n+1$ are equal to $4n(n+1)$. (For $n = 2$ the dimensions of the consequences of the standard identity $s_4 = 0$ and of all multilinear identities of degree 5 are equal respectively to 24 and 29. The difference 5 is equal to the dimension of the vector space spanned by the linearizations $g(x_{\sigma(1)}, \ldots, x_{\sigma(5)})$, $\sigma \in S_5$, of the identity $[[x, y]^2, x] = 0$.)

Concerning the identities of degree $2n + 2$ for $M_n(K)$, all consequences of degree $2n + 2$ of the standard identity $s_{2n} = 0$ were found (in the language of S_{2n+2}-characters) by Benanti and Drensky [BD]. For $n = 5$ and $m = 12$ (the most difficult case handled up till now, but with methods of representation theory of GL_d) this gives $d_F = 8491$. If we work directly with the mulitlinear identities of degree 12, we have $12! = 479\ 001\ 600$ unknowns, $25^{13} \approx 1.49 \cdot 10^{18}$ equations and the number of independent solutions is equal to 8491. Although many of the equations are trivial, the number of the unknowns and the equations which we essentially need (at least $c_{12}(M_5(K)) = 12! - 8491$) is so big that the task is out of reach for present computers.

There is a nice graph theoretical approach to eliminate the trivial equations which was initially created and used for purely theoretic considerations. It is based on arguments used in the Swan proof of the Amitsur–Levitzki theorem [Sw]. Later this approach was used by Formanek [F1] in his construction of central polynomials. For every set of matrix units $\{e_{i_1 j_1}, \ldots, e_{i_k j_k}\} \subset M_n(K)$ we associate an oriented graph with a set of vertices $\{1, \ldots, n\}$ and a set of oriented edges $\{(i_1, j_1), \ldots, (i_k, j_k)\}$. Then the product $e_{i_1 j_1} \cdots e_{i_k j_k}$ is nonzero if and only if $(i_1, j_1), \ldots, (i_k, j_k)$ is a trail of the graph. Hence it is sufficient to consider only those equations (3.2) and only those summands there which correspond to Eulerian graphs and Eulerian trails, respectively. Of course, one may use some symmetries (e.g., to consider the graphs up to an isomorphism) in order to decrease additionally the amount of calculations. Although these graph theoretic arguments cannot bring the number of the equations in the above mentioned case below the bound $12! - 8491$, they are very useful for other considerations.

Instead of considering all multilinear identities of a given degree, it is better to apply methods based on representation theory of the symmetric group and the general linear group and to split the problem into several easier problems.

For example, the calculations for $n = 5$ and $m = 12$ as performed in [BDDK] used representations of $GL_d(K)$. Instead of working with one system with 12! unknowns, as in the direct attempt to handle the multilinear identities, one decomposes the homogeneous component $A_d^{(12)}$ of degree 12 in the free algebra $A_d = K\langle x_1, \ldots, x_d \rangle$ and the polynomial identities of degree 12:

$$A_d^{(12)} = \bigoplus_{\lambda \vdash 12} d_\lambda W_d(\lambda), \quad d_\lambda = \dim M(\lambda),$$

$$A_d^{(12)} \cap T(M_5(K)) = \bigoplus_{\lambda \vdash 12} n_\lambda W_d(\lambda).$$

One searches for polynomial identities in the GL_d-module $d_\lambda W_d(\lambda)$. As a lower bound for the multiplicities n_λ one uses the S_{2n+2}-character of $P_{2n+2} \cap (s_{2n})^T$ obtained in [BD]. In this way, one considers 77 systems with up to 7700 unknowns and finds only sufficiently good bounds for the ranks of the systems. The number of the systems can be decreased to 33, the number of unknowns to 5775 and the computing time to two hours if one uses all the available theoretical information.

We shall complete this chapter with several results on cocharacters of matrices.

The exact cocharacter sequence of $M_n(K)$, $n \geq 2$, is known for $n = 2$ only.

Theorem 3.2.5. (Formanek [F3] and Drensky[Dr3]) *The cocharacter sequence of the T-ideal $T(M_2(K))$ is*

$$\chi_k(M_2(K)) = \sum_{\lambda \vdash k} m_\lambda(M_2(K))\chi_\lambda, \quad k = 0, 1, 2, \ldots,$$

where $\lambda = (\lambda_1, \lambda_2, \lambda_3, \lambda_4)$ and
   (i) $m_{(k)}(M_2(K)) = 1$;
   (ii) $m_{(\lambda_1, \lambda_2)}(M_2(K)) = (\lambda_1 - \lambda_2 + 1)\lambda_2$, *if $\lambda_2 > 0$*;
   (iii) $m_{(\lambda_1, 1, 1, \lambda_4)}(M_2(K)) = \lambda_1(2 - \lambda_4) - 1$;
   (iv) $m_\lambda(M_2(K)) = (\lambda_1 - \lambda_2 + 1)(\lambda_2 - \lambda_3 + 1)(\lambda_3 - \lambda_4 + 1)$ *for all other partitions.*

In 1977, Olsson and Regev [OR] computed the S_{k+1}-character of the multi-linear consequences of degree $k + 1$ of the standard polynomial s_k:

$$\chi_{S_{k+1}}((s_k)^T \cap P_{k+1}) = \chi_{(3,1^{k-2})} + \chi_{(2^2,1^{k-3})} + 3\chi_{(2,1^{k-1})} + \chi_{(1^{k+1})}.$$

The decomposition of Benanti and Drensky [BD] for the consequences of degree $k + 2$ of $s_k = 0$, $k \geq 5$, (obtained in 1998) is:

$$\chi_{S_{k+2}}((s_k)^T \cap P_{k+2}) = 2\chi_{(4,2,1^{k-4})} + 4\chi_{(4,1^{k-2})} + 2\chi_{(3,2^2,1^{k-5})}$$
$$+ 8\chi_{(3,2,1^{k-3})} + 9\chi_{(3,1^{k-1})} + 4\chi_{(2^3,1^{k-4})} + 8\chi_{(2^2,1^{k-2})} + 5\chi_{(2,1^k)} + \chi_{(1^{k+2})}.$$

For more discussions on the topic see [BDDK].

Chapter 4

Central Polynomials for Matrices

4.1 The Approach of Razmyslov

Definition 4.1.1. Let R be an algebra. The polynomial $c(x_1, \ldots, x_m) \in K\langle X \rangle$ is called a *central polynomial* for R if it has no constant term, $c(r_1, \ldots, r_m)$ belongs to the centre of R for all $r_1, \ldots, r_m \in R$, and $c(x_1, \ldots, x_m) = 0$ is not a polynomial identity for R.

Example 4.1.2. By Examples 1.1.5(v), $[[x_1, x_2]^2, x_3] = 0$ is a polynomial identity for $M_2(K)$. Since there exist 2×2 matrices r_1, r_2 such that $[r_1, r_2]^2 \neq 0$ (e.g., $r_1 = e_{12}, r_2 = e_{21}$),

$$c(x_1, x_2) = [x_1, x_2]^2$$

is a central polynomial for the matrix algebra $M_2(K)$.

In 1956 Kaplansky [Ka2], see also the revised version [Ka3] from 1970, gave a list of problems which motivated significant research activity in the subsequent decades. One of his problems was the following.

Problem 4.1.3. (Kaplansky) *Does there exist a multihomogeneous central polynomial for the matrix algebra $M_n(K)$, $n > 2$?*

The answer to the problem of Kaplansky was given in 1972–1973 by Formanek [F1] and Razmyslov [Ra2]. This was very fruitful for PI-theory. Many important results were established or their proofs were simplified using central polynomials for the matrix algebras, see the books by Jacobson [J], Rowen [Rw1] and Formanek [F6].

Here we give the Razmyslov approach to central polynomials with some additional modifications in the concrete constructions. Later, an alternative proof of the existence of central polynomials based on results of Amitsur was given by Kharchenko [Kh]. Combining ideas of Formanek and Razmyslov with other methods, central polynomials for matrices of any size were also constructed by Halpin [Ha1], Giambruno and Valenti [GV]. See also the comments in the next section on the central polynomials constructed by the author with Kasparian [DK2] and Piacentini Cattaneo [DPC] or alone [Dr7].

The method of Razmyslov uses the notion of weak polynomial identities [Ra1, Ra2] and of the Razmyslov transform [Ra2]. See also the book by Razmyslov [Ra5] for account on his methods with applications to many other problems of the theory of associative and Lie algebras and group theory.

Definition 4.1.4. The polynomial $f(x_1, \ldots, x_m) \in K\langle X \rangle$ is called a *weak polynomial identity* for the matrix algebra $M_n(K)$ if $f(a_1, \ldots, a_m) = 0$ for all traceless matrices $a_1, \ldots, a_m$ in $M_n(K)$ (i.e., $a_1, \ldots, a_m$ belong to the Lie subalgebra $sl_n(K)$ of $M_n(K)$). The weak polynomial identity is *essential* if it is not a polynomial identity for $M_n(K)$.

Examples 4.1.5. (i) By Example 1.1.5 (v), if the trace of $a \in M_2(K)$ is equal to 0, then a^2 is a scalar matrix and $[a^2, b] = 0$ for all $b \in M_2(K)$. Hence $[x_1^2, x_2] = 0$ is an essential weak polynomial identity for $M_2(K)$.

(ii) The vector space $sl_n(K)$ is of dimension $n^2 - 1$. Hence every n^2 matrices from $sl_n(K)$ are linearly dependent and the Capelli identity

$$d_{n^2}(x_1, \ldots, x_{n^2}; y_1, \ldots, y_{n^2-1}) = 0$$

vanishes when the n^2 skew-symmetric variables $x_1, \ldots, x_{n^2}$ are replaced by elements from $sl_n(K)$ and the other variables $y_1, \ldots, y_{n^2-1}$ are replaced by arbitrary matrices of $M_n(K)$. Hence $d_{n^2} = 0$ is a weak polynomial identity for $M_n(K)$. It is essential because $d_{n^2} \neq 0$ in $M_n(K)$ by Lemma 3.1.6.

Lemma 4.1.6. (Halpin [Ha1]) *Let S_n act on the set $\{0, 1, \ldots, n-2, n\}$. Then*

$$w(x, y_1, \ldots, y_{n-1}) = \sum_{\sigma \in S_n} (\text{sign } \sigma) x^{\sigma(0)} y_1 x^{\sigma(1)} y_2 \cdots y_{n-2} x^{\sigma(n-2)} y_{n-1} x^{\sigma(n)} = 0$$

is an essential weak polynomial identity for $M_n(K)$ which vanishes for $x \in sl_n(K)$ and $y_1, \ldots, y_{n-1} \in M_n(K)$.

Proof. By the Cayley–Hamilton theorem for any $a \in M_n(K)$

$$a^n - \text{tr}(a)a^{n-1} + \cdots + (-1)^n \det(a)e = 0.$$

If $\text{tr}(a) = 0$, then $e, a, a^2, \ldots, a^{n-2}, a^n$ are linearly dependent and

$$w(a, b_1, \ldots, b_{n-1}) = 0, \quad b_1, \ldots, b_{n-1} \in M_n(K).$$

On the other hand, if $a = \sum_{p=1}^n \rho_p e_{pp}$, then easy calculations give that

$$w(a, e_{12}, e_{23}, \ldots, e_{n-1,n}) = \Delta(\rho_1, \ldots, \rho_n)e_{1n},$$

where

$$\Delta(\rho_1, \ldots, \rho_n) = \begin{vmatrix} 1 & 1 & 1 & \cdots & 1 & 1 \\ \rho_1 & \rho_2 & \rho_3 & \cdots & \rho_{n-1} & \rho_n \\ \rho_1^2 & \rho_2^2 & \rho_3^2 & \cdots & \rho_{n-1}^2 & \rho_n^2 \\ \vdots & \vdots & \vdots & \ddots & \vdots & \vdots \\ \rho_1^{n-2} & \rho_2^{n-2} & \rho_3^{n-2} & \cdots & \rho_{n-1}^{n-2} & \rho_n^{n-2} \\ \rho_1^n & \rho_2^n & \rho_3^n & \cdots & \rho_{n-1}^n & \rho_n^n \end{vmatrix}$$

$$= (\rho_1 + \cdots + \rho_n) \prod_{1 \leq p < q \leq n} (\rho_q - \rho_p).$$

Hence $w(a, e_{12}, e_{23}, \ldots, e_{n-1,n}) \neq 0$ if the eigenvalues of A are pairwise different and its trace is nonzero. Therefore $w(x, y_1, \ldots, y_{n-1}) = 0$ is not an "ordinary" polynomial identity for $M_n(K)$. $\square$

Lemma 4.1.7. *If $f(x_1, \ldots, x_m) = 0$ is a weak polynomial identity for $M_n(K)$, then $f([x_1, x_{m+1}], [x_2, x_{m+2}], \ldots, [x_m, x_{2m}]) = 0$ is an ordinary polynomial identity.*

Proof. Since $\operatorname{tr}([a, b]) = 0$ for any $a, b \in M_n(K)$, we obtain that $f(x_1, \ldots, x_m)$ vanishes on $M_n(K)$ when we replace the variables with commutators. This means that $f([x_1, x_{m+1}], [x_2, x_{m+2}], \ldots, [x_m, x_{2m}]) = 0$ is an ordinary polynomial identity for the matrix algebra $M_n(K)$. $\square$

Many results in PI-theory and its applications (e.g., invariant theory of matrices) involve the properties of trace functions on the algebras. We recall some elementary properties of the traces of matrices.

Lemma 4.1.8. *Any matrix with trace equal to zero in $M_n(K)$ is a sum of at most $n^2 - 1$ commutators.*

Proof. Let us consider the basis of $sl_n(K)$ consisting of e_{ij}, $i \neq j$, and $e_{11} - e_{ii}$, $i = 2, \ldots, n$. The equalities

$$e_{ij} = [e_{ij}, e_{jj}], \; i \neq j, \; \text{and} \; [e_{1i}, e_{i1}] = e_{11} - e_{ii}$$

give that the basis elements of $sl_n(K)$ are commutators. Since $\alpha[x, y] = [\alpha x, y]$, $\alpha \in K$, the lemma follows immediately. $\square$

Lemma 4.1.9. *Let us define a symmetric bilinear form on the vector space $V = M_n(K)$ by*

$$\langle a, b \rangle = \operatorname{tr}(ab), \quad a, b \in M_n(K).$$

This form is nondegenerate, i.e., $\langle u, V \rangle = 0$ implies $u = 0$.

Proof. The lemma follows immediately from the observation that we always can multiply the nonzero matrix $u \in M_n(K)$ by a suitable matrix $a \in M_n(K)$ in order to obtain $\operatorname{tr}(ua) \neq 0$. $\square$

Now we introduce the Razmyslov transform and prove his lemma. These are the main tools in his construction of central polynomials.

Definition 4.1.10. Let $f(x, y_1, \ldots, y_m)$ be a polynomial in $K \langle x, y_1, \ldots, y_m \rangle$ which is linear in the variable x. Writing f in the form $f = \sum g_i x h_i$, where $g_i, h_i \in K \langle y_1, \ldots, y_m \rangle$, the *Razmyslov transform* of f is the polynomial

$$f^*(x, y_1, \ldots, y_m) = \sum h_i x g_i.$$

For example, if

$$f(x, y_1, y_2) = [xy_1 + y_1 x, y_2] = 1 \cdot x \cdot y_1 y_2 + y_1 \cdot x \cdot y_2 - y_2 \cdot x \cdot y_1 - y_2 y_1 \cdot x \cdot 1,$$

then
$$f^*(x, y_1, y_2) = y_1 y_2 \cdot x \cdot 1 + y_2 \cdot x \cdot y_1 - y_1 \cdot x \cdot y_2 - 1 \cdot x \cdot y_2 y_1$$
$$= [y_2, x] y_1 + y_1 [y_2, x].$$

Theorem 4.1.11. (Lemma of Razmyslov [Ra2]) *Let K be any field and let the polynomial*
$$f = f(x, y_1, \dots, y_m) \in K\langle x, y_1, \dots, y_m \rangle$$
be homogeneous of first degree in x and let $f^* = f^*(x, y_1, \dots, y_m)$ be the polynomial obtained by the Razmyslov transform. Then:

(i) *If $f = 0$ is a polynomial identity for $M_n(K)$, then $f^* = 0$ is also a polynomial identity.*

(ii) *If $f = 0$ is a weak polynomial identity for $M_n(K)$, then f^* is either a weak polynomial identity or has only scalar values on $sl_n(K)$.*

(iii) *If $f = 0$ is an essential weak polynomial identity for $M_n(K)$ such that*
$$f([x, z], y_1, \dots, y_m) = 0$$
is a polynomial identity for $M_n(K)$, then f^* is a central polynomial for $M_n(K)$.

Proof. (i) Let
$$f = \sum g_i x h_i, \quad g_i, h_i \in K\langle y_1, \dots, y_m \rangle.$$
Then, by Lemma 4.1.9, the trace is nondegenerate and $f = 0$ is a polynomial identity for $M_n(K)$ if and only if $\text{tr}(fz) = 0$ is a trace identity (i.e., vanishes on $M_n(K)$). Since $\text{tr}(uv) = \text{tr}(vu)$, we obtain
$$\text{tr}(f(x, y_1, \dots, y_m)z) = \text{tr}(fz) = \sum \text{tr}(g_i x h_i z)$$
$$= \sum \text{tr}(h_i z g_i x) = \text{tr}(f^*(z, y_1, \dots, y_m)x).$$

Hence $f(x, y_1, \dots, y_m) = 0$ is a polynomial identity for $M_n(K)$ if and only if $\text{tr}(f^*(z, y_1, \dots, y_m)x) = 0$ is a trace identity, which, by Lemma 4.1.9 again, is equivalent to the fact that $f^*(z, y_1, \dots, y_m) = 0$ is a polynomial identity.

(ii) Let
$$f(x, y_1, \dots, y_m) = \sum g_i(y_1, \dots, y_m) x h_i(y_1, \dots, y_m) = 0$$
be a weak polynomial identity for $M_n(K)$. In f we replace x by $[x, z]$, each y_j by a sum $\sum [y_{jk}, z_{jk}]$ of $n^2 - 1$ commutators $[y_{jk}, z_{jk}]$, $k = 1, \dots, n^2 - 1$, and obtain that
$$\tilde{f} = f\left([x, z], \sum_{k=1}^{n^2-1} [y_{1k}, z_{1k}], \dots, \sum_{k=1}^{n^2-1} [y_{mk}, z_{mk}]\right) = 0$$
is a polynomial identity for $M_n(K)$. Since
$$\tilde{f} = \sum \tilde{g}_i [x, z] \tilde{h}_i = \sum (\tilde{g}_i x z \tilde{h}_i - \tilde{g}_i z x \tilde{h}_i),$$

where

$$\tilde{g}_i = g_i \left(\sum [y_{1k}, z_{1k}], \ldots, \sum [y_{mk}, z_{mk}] \right),$$

$$\tilde{h}_i = h_i \left(\sum [y_{1k}, z_{1k}], \ldots, \sum [y_{mk}, z_{mk}] \right),$$

by (i) we obtain that

$$\tilde{f}^* = \sum (z \tilde{h}_i x \tilde{g}_i - \tilde{h}_i x \tilde{g}_i z)$$
$$= \left[z, \sum (\tilde{h}_i x \tilde{g}_i - \tilde{h}_i x \tilde{g}_i) \right] = \left[z, f^* \left(x, \sum [y_{1k}, z_{1k}], \ldots, \sum [y_{mk}, z_{mk}] \right) \right] = 0$$

is also a polynomial identity for $M_n(K)$. Applying Lemma 4.1.8 we derive that $[z, f^*(x, y_1, \ldots, y_m)]$ vanishes when we replace $y_1, \ldots, y_m$ by elements of $sl_n(K)$ and x, z by any elements of $M_n(K)$. Hence $f^*(x, y_1, \ldots, y_m)$ either vanishes or takes scalar values when $y_1, \ldots, y_m$ are replaced by traceless $n \times n$ matrices and x by any $n \times n$ matrix. (If the weak polynomial identity $f = 0$ is multilinear it is sufficient to replace y_j by $[y_j, z_j]$. We needed the sum in order to handle the case of an arbitrary field.)

(iii) Let $f = 0$ be an essential weak polynomial identity for $M_n(K)$ such that $f([x, z], y_1, \ldots, y_m) = 0$ is a polynomial identity. As in (ii),

$$f^*([x, z], y_1, \ldots, y_m) = [z, f^*(x, y_1, \ldots, y_m)] = 0$$

is a polynomial identity. Hence $f^*(x, y_1, \ldots, y_m)$ has scalar values on $M_n(K)$. Since $f = 0$ is not a polynomial identity, we conclude that $f^*(r, r_1, \ldots, r_m) \neq 0$ for some $r, r_1, \ldots, r_m \in M_n(K)$ and this means that f^* is a nontrivial central polynomial. □

Theorem 4.1.12. (Razmyslov [Ra2]) *Let* $d_{n^2}(x_1, \ldots, x_{n^2}; y_1, \ldots, y_{n^2-1})$ *be the Capelli polynomial in* n^2 *skew-symmetric variables and let*

$$f = f(x, z_1, \ldots, z_{2n^2-2}, y_1, \ldots, y_{n^2-1})$$
$$= d_{n^2}(x, [z_1, z_2], \ldots, [z_{2n^2-3}, z_{2n^2-2}]; y_1, \ldots, y_{n^2-1}).$$

The Razmyslov transform f^* *applied to* f *is a multilinear central polynomial for* $M_n(K)$ *over any field* K.

Proof. By Examples 4.1.5 (ii), the Capelli identity $d_{n^2} = 0$ is a weak polynomial identity for $M_n(K)$. Since the commutators of the elements of $M_n(K)$ are in $sl_n(K)$, obviously $f = 0$ is also a weak polynomial identity which vanishes when we replace x by an element of $sl_n(K)$ and $y_1, \ldots, y_{n^2-1}$ and $z_1, \ldots, z_{2n^2-2}$ by any matrices. Hence,

$$f([x, t], z_1, \ldots, z_{2n^2-2}, y_1, \ldots, y_{n^2-1}) = 0$$

is a polynomial identity. In order to apply the lemma of Razmyslov, it is sufficient to show that

$$f(x, z_1, \ldots, z_{2n^2-2}, y_1, \ldots, y_{n^2-1}) = 0$$

does not vanish on $M_n(K)$. By Lemma 3.1.6, $d_{n^2}(x_1,\ldots,x_{n^2};y_1,\ldots,y_{n^2-1})=0$ is not a polynomial identity of $M_n(K)$. Since d_{n^2} is skew-symmetric in the x's, it does not vanish if we replace $x_1,\ldots,x_{n^2}$ by the elements of any basis $\{r_1,\ldots,r_{n^2}\}$ of $M_n(K)$ and $y_1,\ldots,y_{n^2-1}$ by suitable $n\times n$ matrices $s_1,\ldots,s_{n^2-1}$. We shall fix the basis

$$\{r_1,\ldots,r_{n^2}\}=\{e_{11},e_{11}-e_{ii},e_{jk}\mid i=2,\ldots,n,\quad j\neq k,\quad j,k=1,\ldots,n\},$$

where we denote e_{11} by r_1 and the other elements by $r_2,\ldots,r_{n^2}$. Then

$$d_{n^2}(r_1,\ldots,r_{n^2};s_1,\ldots,s_{n^2-1})\neq 0.$$

Since $r_2,\ldots,r_{n^2}$ are equal to commutators, $r_p=[a_p,b_p]$, $p=2,\ldots,n^2$, we obtain that

$$d_{n^2}(r_1,\ldots,r_{n^2};s_1,\ldots,s_{n^2-1})=f(r_1,[a_2,b_2],\ldots,[a_{n^2},b_{n^2}],s_1,\ldots,s_{n^2-1})\neq 0,$$

and this completes the proof. $\qquad\qquad\qquad\qquad\qquad\qquad\qquad\qquad\qquad\square$

In order to obtain central polynomials by the method of Razmyslov, we need an essential weak polynomial identity $w(x_1,\ldots,x_m)=0$ for $M_n(K)$. Then $w(x_1,\ldots,x_m)=0$ is not a polynomial identity and $w([y_1,z_1],\ldots,[y_m,z_m])=0$ is. Hence there is a p such that $f=w([y_1,z_1],\ldots,[y_p,z_p],x_{p+1},\ldots,x_m)$ does not vanish on $M_n(K)$ and

$$w([y_1,z_1],\ldots,[y_p,z_p],[y_{p+1},z_{p+1}],x_{p+2},\ldots,x_m)=0$$

is a polynomial identity. Then, applying the Razmyslov transform to f with respect to x_{p+1} we obtain a central polynomial.

Example 4.1.13. By Example 4.1.5 (i),

$$w(x,y)=[x^2,y]=0$$

is an essential weak polynomial identity for $M_2(K)$. Its linearization

$$w_1(x,y,z)=[xz+zx,y]=xzy+zxy-yxz-yzx=0$$

is also a weak identity. Replacing in $w_1(x,y,[u,v])$ suitable elements of $M_2(K)$ we obtain that

$$f(x,y,u,v)=w_1(x,y,[u,v])$$

is not an ordinary identity for $M_2(K)$. For example,

$$w_1(e_{11}+e_{22},e_{22},[e_{11},e_{12}])=2[e_{12},e_{22}]=2e_{12}\neq 0.$$

On the other hand, $w_1([x,t],y,[u,v])=0$ is a polynomial identity. Hence the Razmyslov transform f^* with respect to x applied on $f(x,y,u,v)$ is a central polynomial. We express f in the form

$$f=[x[u,v]+[u,v]x,y]=x[u,v]y+[u,v]xy-yx[u,v]-y[u,v]x$$

and the obtained central polynomial is

$$c(x, y, u, v) = f^* = [u, v]yx + yx[u, v] - [u, v]xy - xy[u, v]$$
$$= [u, v][y, x] + [y, x][u, v].$$

If we substitute $v = x$ and $u = y$, we obtain

$$c(x, y, y, x) = 2[x, y]^2$$

which is the known central polynomial for $M_2(K)$.

Using the method of Razmyslov, Halpin [Ha1] constructed another central polynomial.

Theorem 4.1.14. (Halpin [Ha1]) *Let S_n act on $\{0, 1, 2, \ldots, n - 2, n\}$ and let*

$$w_1(x, Z, Y) = w_1(x, z_1, \ldots, z_m, y_1, \ldots, y_{n-1}), \ m = \frac{1}{2}n(n-1),$$

be the complete linearization in x of

$$w(x, Y) = \sum_{\sigma \in S_n} (\text{sign } \sigma) x^{\sigma(0)} y_1 x^{\sigma(1)} y_2 \cdots y_{n-2} x^{\sigma(n-2)} y_{n-1} x^{\sigma(n)},$$

i.e., the multilinear component of $w(x + z_1 + \cdots + z_m, y_1, \ldots, y_{n-1})$. If

$$f(x, U, V, Y) = w_1(x, [u_1, v_1], \ldots, [u_m, v_m], y_1, \ldots, y_{n-1}),$$

then the polynomial $f^*(x, U, V, Y)$ obtained by applying the Razmyslov transform to $f(x, U, V, Y)$ is central for $M_n(K)$.

Proof. By Lemma 4.1.6, $w_1(x, Z, Y) = 0$ is a weak polynomial identity for $M_n(K)$ which vanishes when we replace $x, z_1, \ldots, z_m$ by elements from $sl_n(K)$ and $y_1, \ldots, y_{n-1}$ by any matrices. Hence $f([x, t], U, V, Y) = 0$ is an ordinary polynomial identity. If we show that $f(x, U, V, Y)$ has a nonzero evaluation on $M_n(K)$, then the lemma of Razmyslov gives that $f^*(x, U, V, Y)$ is a central polynomial. Since $w_1(x, Z, Y)$ is of degree m with respect to $z_1, \ldots, z_m$, we obtain that

$$w_1(x, z, \ldots, z, y_1, \ldots, y_{n-1}) = m! w_2(x, z, y_1, \ldots, y_{n-1}),$$

where $w_2(x, z, Y)$ is equal to the linear component in x of $w(x + z, Y)$. We shall show that $w_2(a, c, b_1, \ldots, b_{n-1}) \neq 0$ for

$$a = e, \quad b_1 = e_{12}, b_2 = e_{23}, \ldots, b_{n-1} = e_{n-1,n},$$

$$c = \sum_{p=2}^{n} \tau_p(-e_{11} + e_{pp}) = \left[\sum_{q=2}^{n} e_{q,q-1}, \sum_{p=2}^{n} (\tau_p + \tau_{p+1} + \cdots + \tau_n)e_{p-1,p}\right].$$

Since $c = [c_1, c_2]$ is a commutator, this will give that

$$w_2(a, [c_1, c_2], b_1, \ldots, b_{n-1}) \neq 0$$

and hence $f(x, U, V, Y)$ will not be a polynomial identity for $M_n(K)$. Since

$$a + c = \rho_1 e_{11} + \cdots + \rho_n e_{nn},$$

where

$$\rho_1 = 1 - (\tau_2 + \cdots + \tau_n), \quad \rho_p = 1 + \tau_p, \quad p = 2, \ldots, n,$$

we have

$$\rho_1 + \cdots + \rho_n = n,$$

$$\rho_p - \rho_1 = (\tau_2 + \cdots + \tau_n) + \tau_p, \quad p = 2, \ldots, n,$$

$$\rho_q - \rho_p = \tau_q - \tau_p, \quad 2 \leq p < q \leq n.$$

We use the proof of Lemma 4.1.6 and obtain that

$$w(a + c, b_1, \ldots, b_{n-1}) = n e_{1n} \prod_{p=2}^{n} ((\tau_2 + \cdots + \tau_n) + \tau_p) \prod_{2 \leq p < q \leq n} (\tau_q - \tau_p)$$

and this is not zero for suitably chosen $\tau_2, \ldots, \tau_n$. Let

$$w(a + c, b_1, \ldots, b_{n-1}) = (h_0 + h_1 + \cdots + h_m + h_{m+1}) e_{1n},$$

where $h_i e_{1n}$ is the homogeneous component of degree i with respect to a. Clearly, h_i is a homogeneous polynomial of degree $m + 1 - i$ in $\tau_2, \ldots, \tau_n$. Hence, the linear component in a of $w(a + c, b_1, \ldots, b_{n-1})$ is equal to $w(a + c, b_1, \ldots, b_{n-1})$ itself and this gives that $w_2(x, z, Y) \neq 0$ in $M_n(K)$. $\qquad \square$

Remark 4.1.15. The problem of the existence of central polynomials is much easier in the case of finite fields. Latyshev and Shmelkin [LS] constructed a central polynomial in one variable for $M_n(\mathbb{F}_q)$, where $\mathbb{F}_q$ is the field with q elements. If $p(x)$ is an irreducible polynomial of degree n in $\mathbb{F}_q[x]$, then

$$c(x) = \prod_{m=1}^{n-1} (x^{q^m} - x)^{(q^n - 1)n} \left(\frac{x^{q^n} - x}{p(x)} \right)^{(q^n - 1)n}$$

is a nontrivial central polynomial for $M_n(\mathbb{F}_q)$.

4.2   More Central Polynomials

The description of the central polynomials is known for 2×2 matrices only (and when char $K = 0$). Formanek [F3] gave the complete quantitative information in terms of Hilbert series and S_m-cocharacters. Okhitin [Ok2] established the following result.

Theorem 4.2.1. (Okhitin [Ok2]) *Modulo the polynomial identities for $M_2(K)$, any central polynomial for $M_2(K)$ is a linear combination of polynomials of the form*

$$[u_1, u_2][u_3, u_4] + [u_3, u_4][u_1, u_2], \quad u_1, u_2, u_3, u_4 \in K\langle X \rangle.$$

One can see that the central polynomial

$$[x_1, x_2][x_3, x_4] + [x_3, x_4][x_1, x_2]$$

is a linear combination of the linearizations of the "trivial" central polynomial $[x, y]^2$ and the standard polynomial $s_4(x_1, x_2, x_3, x_4)$. In other words, the central polynomials of $M_2(K)$ "follow" from the central polynomial $[x, y]^2$ and the standard identity $s_4 = 0$.

For different purposes one needs different central polynomials. In particular, it is important to know the minimal degree of the central polynomials.

Problem 4.2.2. (See Formanek [F2, F6]) *Find the minimal degree of the central polynomials for $M_n(K)$, char $K = 0$. Find the minimal degree of the central polynomials in two variables.*

The central polynomials of Formanek [F1] are of degree n^2. The original polynomials of Razmyslov [Ra2], see Theorem 4.1.12, are of degree $3n^2 - 2$, but using other weak polynomial identities, Halpin [Ha1], see Theorem 4.1.14, also reduced the degree to n^2. One may conjecture that the minimal degree of the central polynomials for $M_n(K)$ is equal to n^2, and this is true for $n = 1, 2$. For $n = 3$ Drensky and Kasparian [DK2] constructed a central polynomial of degree 8. They used ideas of the Rosset proof of the Amitsur–Levitzki theorem [Ro]. They also proved [DK1] that $M_3(K)$ has no central polynomials of degree 7. The very computational proof, done by hand, combines calculations with proper polynomial identities and techniques of representation theory of the general linear group.

Although this is not a reason for the following conjecture by Formanek, the only quadratic function $p(n)$ with $p(1) = 1$, $p(2) = 4$ and $p(3) = 8$ is $\frac{1}{2}(n^2 + 3n - 2)$.

Conjecture 4.2.3. (Formanek [F6]) *The minimal degree of the central polynomials for $M_n(K)$ over a field K of characteristic 0 is*

$$\mathrm{mindeg}(M_n(K)) = \frac{1}{2}(n^2 + 3n - 2).$$

It is important that the conjecture of Formanek is in accordance with some other conjectures in the theory of PI-algebras and in particular with the conjecture of Kuzmin about the class of nilpotency in the Nagata-Higman theorem, see Conjecture 6.2.2 below.

Drensky and Rashkova [DR] found all weak polynomial identities of degree 6 for $M_3(K)$. Besides the weak identity of Halpin (which is of degree 6 for $M_3(K)$), they found a new weak identity which gives rise to a central polynomial of degree 8 and explained from this point of view the central polynomials of minimal degree for $M_3(K)$. Drensky and Piacentini Cattaneo [DPC] found a new central polynomial of degree 13 for $M_4(K)$. Since $\frac{1}{2}(4^2+3.4-2)=13$ this agrees with the conjecture of Formanek. (Unfortunately we do not know whether $M_4(K)$ has central polynomials of degree 12.) The construction uses a weak polynomial identity of degree 9 and combines the methods of Formanek and Razmyslov. The result was generalized by Drensky [Dr7] who constructed central polynomials of degree $(n-1)^2+4$ for all $M_n(K)$, $n>2$. The central polynomials in [DR, DPC, Dr7] use the essential weak polynomial identity

$$
\begin{aligned}
w(x,y_1,\ldots,y_n) &= s_{2n-2}(x,x^2,\ldots,x^{n-3},x^n,y_1,\ldots,y_n) \\
&+ \sum_{i=1}^{n} x s_{2n-2}(x,x^2,\ldots,x^{n-3},x^{n-2},y_1,\ldots,y_ix,\ldots,y_n) \\
&+ \sum_{1\le i<j\le n} s_{2n-2}(x,x^2,\ldots,x^{n-3},x^{n-2},y_1,\ldots,y_ix,\ldots,y_jx,\ldots,y_n)=0
\end{aligned}
$$

for $M_n(K)$ which vanishes when x is replaced by an element of $sl_n(K)$ and $y_1,\ldots,y_n$ are any matrices of $M_n(K)$. Then the method of Razmyslov gives central polynomials of degree $(n-1)^2+4$. For $n=3$ and $n=4$ this degree coincides with the degree conjectured by Formanek.

The existence of a weak identity of third degree for $M_2(K)$ gives rise to a central polynomial of fourth degree. Similarly, for the construction of central polynomials respectively of degree 8 and 13 for the algebras $M_3(K)$ and $M_4(K)$ we may use weak identities respectively of degree 6 and 9, which is a big advantage from the computational point of view. We refer to the survey article [BDDK] which, among the other topics, deals with computational aspects of central polynomials and weak polynomial identities.

Problem 4.2.4. *Find the minimal degree of the essential weak polynomial identities for $M_n(K)$, char $K=0$. Find weak polynomial identities of low degree which produce central polynomials of low degree.*

The answer is known for $n=2$ where the theorem of Razmyslov [Ra1] shows that all weak polynomial identities are consequences of $[x_1^2,x_2]=0$ and for $n=3$, when the minimal degree of the weak polynomial identities is equal to 6, see [DR].

Chapter 5

Invariant Theory of Matrices

5.1 A Background on Invariant Theory

Let U_m be a vector space with basis $\{u_1, \ldots, u_m\}$ and let $S(U_m)$ be the symmetric algebra of U_m over K. This is the polynomial algebra $K[U_m] = K[u_1, \ldots, u_m]$ with m variables $u_1, \ldots, u_m$. The algebra $S(U_m)$ is naturally graded. Its homogeneous component of degree p is the p-th symmetric power $S^p(U_m)$ of U_m, i.e., the vector space spanned by the monomials of degree p. The general linear group $GL(U_m) \cong GL_m(K)$ acts on the vector space U_m and this action is extended diagonally to the group of homogeneous automorphisms of $K[U_m]$:

$$g(f(u_1, \ldots, u_m)) = f(g(u_1), \ldots, g(u_m)), \quad g \in GL_m(K), \ f \in K[U_m].$$

For a subgroup G of $GL_m(K)$, the set

$$K[U_m]^G = \{f \in K[U_m] \mid g(f) = f \quad \text{for all } g \in G\}$$

is a subalgebra of $K[U_m]$ called the *algebra of invariants* of G. The algebra of invariants inherits the grading of $K[U_m]$:

$$K[U_m]^G = K \oplus U_m^G \oplus S^2(U_m)^G \oplus S^3(U_m)^G \oplus \cdots.$$

The formal power series

$$H(K[U_m]^G, t) = 1 + \dim U_m^G \cdot t + \dim S^2(U_m)^G \cdot t^2 + \dim S^3(U_m)^G \cdot t^3 + \cdots$$

is called the Hilbert (or Poincaré) series of the algebra of invariants.

There are several problems related with the invariants of a group G.

Problem 5.1.1. *Is the algebra of invariants $K[U_m]^G$ finitely generated for any subgroup G of $GL_m(K)$?*

In his famous lecture *"Mathematische Probleme"* given at the International Congress of Mathematicians held in 1900 in Paris, Hilbert [Hi2] asked his more general 14-th Problem which was mainly motivated by the above Problem 5.1.1. For arbitrary groups, the answer is negative, see Nagata [Na2]. See also the survey by Freudenburg [Fr] on different counterexamples to Hilbert's 14th problem.

Problem 5.1.1 has a positive solution for large classes of groups. This holds for finite groups and reductive groups. In particular, the algebra of invariants is finitely generated when the subgroup G of $GL_m(K)$ is isomorphic to $GL_p(K)$, to the special linear group

$$SL_p(K) = \{g \in GL_p(K) \mid \det(g) = 1\},$$

to all semisimple algebraic groups. Below we give the theorem of Emmy Noether and a simplified version of a result of Nagata which extends a theorem of Hilbert.

Theorem 5.1.2. (Endlichkeitssatz, Emmy Noether, [No]) *Let G be a finite subgroup of $GL_m(K)$. Then the algebra of invariants $K[u_1, \ldots, u_m]^G$ is finitely generated. It has a system of generators $f_1, \ldots, f_p$, where every f_i is a homogeneous polynomial whose degree is bounded by the order $|G|$ of the group G.*

Theorem 5.1.3. (Hilbert–Nagata, see, e.g., [DC]) *Let G be a subgroup of $GL_m(K)$ such that any finite dimensional rational representation of G is completely reducible. Then the algebra of invariants $K[u_1, \ldots, u_m]^G$ is finitely generated.*

Let us recall Hilbert's Basissatz [Hi1]:

Theorem 5.1.4. *Every commutative algebra S generated by a finite set $\{f_1, \ldots, f_p\}$ can be defined by a finite system of relations, i.e., there exists a finite system of polynomials*

$$r_1(y_1, \ldots, y_p), \ldots, r_q(y_1, \ldots, y_p) \in K[y_1, \ldots, y_p]$$

such that the kernel of the canonical homomorphism

$$\nu : K[y_1, \ldots, y_p] \longrightarrow S, \quad \text{where} \quad \nu(y_j) = f_j, \ j = 1, \ldots, p,$$

is generated as an ideal by $r_1(y_1, \ldots, y_p), \ldots, r_q(y_1, \ldots, y_p)$.

Hence, if the algebra $K[U_m]^G$ is finitely generated, it is also *finitely presented*, i.e., has a finite system of generators with a finite number of defining relations between them. Traditionally a result giving the explicit generators of the algebra of invariants of a group $G \subset GL_m(K)$ is called the *first fundamental theorem of the invariant theory* of G and a result describing the relations between the generators — the *second fundamental theorem*.

For example, if the symmetric group S_m acts canonically on the basis of U_m by $\sigma(u_i) = u_{\sigma(i)}$, $\sigma \in S_m$, $i = 1, \ldots, m$, the first fundamental theorem states that the elementary symmetric functions $e_i(u_1, \ldots, u_m)$, $i = 1, \ldots, m$, generate the algebra of symmetric polynomials in m variables and the second fundamental theorem says that there are no relations between them, i.e., the elementary symmetric functions are algebraically independent.

The Molien formula gives an expression of the Hilbert series of the algebra of invariants for finite groups.

Theorem 5.1.5. (Molien formula [Mo]) *The Hilbert series of the algebra of invariants of a finite subgroup G of $GL_m(K)$ is*

$$H(K[U_m]^G, t) = \frac{1}{|G|} \sum_{g \in G} \frac{1}{\det(1 - gt)}.$$

In the case of infinite groups G the Molien formula has no formal sense. Nevertheless, if G is compact, one can define Haar measure on G, replace the sum with an integral and obtain the Weyl formula for the Hilbert series of the algebra of invariants, see [We2].

5.2 The Generic Trace Algebra

We shall introduce several objects related both with invariant theory and with the generic matrix algebra R_n. Recall that the generic $n \times n$ matrices are the matrices

$$y_i = \sum_{p,q=1}^{n} y_{pq}^{(i)} e_{pq}, \quad i = 1, 2, \ldots,$$

with entries from the polynomial algebra

$$\Omega_n = K[y_{pq}^{(i)} \mid p, q = 1, \ldots, n, i = 1, 2, \ldots].$$

Clearly, R_n is a K-subalgebra of $M_n(\Omega_n) \cong \Omega_n \otimes M_n(K)$. The first object is the *pure trace algebra* C_n generated as a unitary subalgebra of Ω_n by all traces of products of generic matrices

$$\mathrm{tr}(y_{i_1} \cdots y_{i_m}), \quad i_j = 1, 2, \ldots, \quad m = 1, 2, \ldots.$$

Identifying the identity matrix e with 1, we may assume that $\Omega_n \subset M_n(\Omega_n)$ and define the second object, the *mixed trace algebra* T_n. It is generated by the generic matrices $y_1, y_2, \ldots$ and the elements of C_n.

We denote by R_{nd} and Ω_{nd} the subalgebras of R_n and Ω_n generated, respectively, by $y_1, \ldots, y_d$ and $y_{pq}^{(i)}$, $p, q = 1, \ldots, n$, $i = 1, \ldots, d$. The corresponding subalgebras of C_n and T_n will be, respectively, C_{nd} and T_{nd}.

Theorem 5.2.1. (See, e.g., [F6] Theorem 22, p. 19 and Proposition 43, p. 43) *Let* $n, d \geq 2$.

(i) *The algebra T_{nd} has no zero divisors.*
(ii) *The algebra C_{nd} coincides with the centre of T_{nd}.*
(iii) *The centre of R_{nd} is a subalgebra of C_{nd} with the same quotient field as C_{nd}.*

Let us consider the dn^2-dimensional vector space U with basis the variables

$$y_{pq}^{(i)}, \quad p, q = 1, \ldots, n, \quad i = 1, \ldots, d.$$

Clearly, $K[U] = \Omega_{nd}$. We define an action of $GL_n(K)$ on U, and hence on Ω_{nd} as follows. Let y_i be the i-th generic $n \times n$ matrix and let $g \in GL_n(K)$. Then

$$g * y_i = g y_i g^{-1} = g \left(\sum_{p,q=1}^{n} y_{pq}^{(i)} e_{pq} \right) g^{-1} = \sum_{p,q=1}^{n} z_{pq}^{(i)} e_{pq}$$

for some $z_{pq}^{(i)}$ which are linear combinations of the $y_{pq}^{(i)}$'s. Then we define

$$g * y_{pq}^{(i)} = z_{pq}^{(i)}, \quad p, q = 1, \ldots, n, \quad i = 1, \ldots, d.$$

The elements of the algebra of invariants $\Omega_{nd}^{GL_n(K)}$ under this action are called the *invariants* under the action of $GL_n(K)$ by *simultaneous conjugation* of d matrices of size $n \times n$. Similarly, we define the action of $GL_n(K)$ on the polynomial algebra Ω_n in countably many variables $y_{pq}^{(i)}$, $p, q = 1, \ldots, n$, $i = 1, 2, \ldots$, and consider the algebra of invariants $\Omega_n^{GL_n(K)}$.

Many basic properties of $\Omega_{nd}^{GL_n(K)}$ can be derived directly from the fact that the group $GL_n(K)$ is reductive. For example, by Theorem 5.1.3 one obtains that the algebra $\Omega_{nd}^{GL_n(K)}$ is noetherian. We collect some of these properties in the following theorem. The sketch of its proof with additional comments can be found, e.g., in [F3], p. 42.

Theorem 5.2.2. *The algebra $\Omega_{nd}^{GL_n(K)}$ is finitely generated. It is a Cohen–Macaulay and even Gorenstein unique factorization domain.*

The concrete form of the algebra of invariants $\Omega_n^{GL_n(K)}$ (and hence of its subalgebras $\Omega_{nd}^{GL_n(K)}$) is given by the following theorem:

Theorem 5.2.3. (First Fundamental Theorem of Matrix Invariants) *The algebra of matrix invariants $\Omega_n^{GL_n(K)}$ coincides with the pure trace algebra C_n generated by all traces of products of generic matrices*

$$\mathrm{tr}(y_{i_1} \cdots y_{i_m}), \quad i_j = 1, 2, \ldots, \quad m = 1, 2, \ldots.$$

It is difficult to judge about the priority in the proof of this theorem. (See the history of generic matrices in the paper by Formanek [F7].) The proof follows from general facts on invariant theory of $GL_n(K)$ and can be found in the book by Gurevich [Gu] and in the papers by Sibirskii [Si] and Procesi [Pr3]. Although it was stated as well known by Kirillov [Ki], it seems that the understanding of the importance of the theorem is a result of its rediscovery by Procesi [Pr3].

Of course, the algebra C_n is not finitely generated. In the next chapter we shall use purely PI-techniques to give an explicit upper bound for the degree of the traces in the natural set of generators of C_n. (It is not clear at all that the algebra C_n can be generated by traces of bounded degree.)

A minimal set of generators for the invariants of 2×2 matrices was found by Sibirskii [Si]. We shall discuss invariant theory of 2×2 matrices in the next section. Abeasis and Pittaluga [AP] suggested an algorithm for finding a minimal set of generators for C_{nd}. Combining computers with calculations by hand they successfully aplied this algorithm to the case of 3×3 matrices. Teranishi [T1, T2] found explicit systems of generators of C_{32} and C_{42}. In particular, the algebra C_{32} of the invariants of two 3×3 matrices has the following system of generators

$$\mathrm{tr}(x), \mathrm{tr}(y), \mathrm{tr}(x^2), \mathrm{tr}(xy), \mathrm{tr}(y^2),$$
$$\mathrm{tr}(x^3), \mathrm{tr}(x^2 y), \mathrm{tr}(xy^2), \mathrm{tr}(y^3), \mathrm{tr}(x^2 y^2), \mathrm{tr}(x^2 y^2 xy), \tag{5.1}$$

where x, y are generic 3×3 matrices.

The second fundamental theorem or the problem for the description of the defining relations of the algebra of $n \times n$ matrix invariants has two aspects. One of them is to handle the case of the algebra C_{nd} for fixed d. The other is to consider the multilinear invariants only. The reason is that if we know the relations between the multilinear invariants, then we know all the relations. The arguments are similar to those that every T-ideal $T(R)$ of the free algebra $K\langle X \rangle$ is generated as a T-ideal by its multilinear elements. The second aspect is completely solved by the theorem of Razmyslov–Procesi [Ra3, Pr3]. We shall state a version of the result only. We need some preliminary discussions. For a fixed m we write the permutations $\sigma \in S_m$ as products of disjoint cycles

$$\sigma = (i_1 \ldots i_p) \cdots (j_1 \ldots j_q),$$

including also the 1-cycles, so that each integer $1, \ldots, m$ occures exactly once. We define the *associated trace function*

$$\mathrm{tr}_\sigma(x_1, \ldots, x_n) = \mathrm{tr}(x_{i_1} \cdots x_{i_p}) \cdots \mathrm{tr}(x_{j_1} \cdots x_{j_q}).$$

We also assume that for $m \leq k$ the symmetric group S_m acts on $1, \ldots, m$ and leaves invariant $m + 1, \ldots, k$, i.e., S_m is canonically embedded into S_k.

Theorem 5.2.4. (Second Fundamental Theorem of Matrix Invariants [Ra3, Pr3])
   (i) *Let*

$$f(x_1, \ldots, x_m) = \sum_{\sigma \in S_m} \alpha_\sigma \mathrm{tr}_\sigma(x_1, \ldots, x_m), \quad \alpha_\sigma \in K,$$

be a multilinear trace polynomial of degree m. Then $f = 0$ is a trace identity for the $n \times n$ matrix algebra, i.e., $f(a_1, \ldots, a_m) = 0$ for all $a_1, \ldots, a_m \in M_n(K)$, if and only if

$$\sum_{\sigma \in S_m} \alpha_\sigma \sigma$$

belongs to the two-sided ideal $J(n, m)$ of the group algebra KS_m generated by the element

$$\sum_{\sigma \in S_{n+1}} (\mathrm{sign}\ \sigma)\sigma.$$

   (ii) *The ideal $J(n, m)$ of KS_m is a direct sum of the minimal two-sided ideals of KS_m corresponding to the partitions $\lambda = (\lambda_1, \ldots, \lambda_m)$ with at least $n + 1$ parts, i.e., satisfying $\lambda_{n+1} \neq 0$.*

The *fundamental trace identity*

$$\sum_{\sigma \in S_{n+1}} (\mathrm{sign}\ \sigma)\mathrm{tr}_\sigma(x_1, \ldots, x_{n+1}) = 0$$

is actually the linearization of the Cayley–Hamilton theorem. We illustrate this for 2×2 matrices. As in the proof of Razmyslov of the Amitsur–Levitzki theorem, the Cayley–Hamilton theorem for 2×2 matrices can be rewritten in the form of the mixed trace identity

$$c(x) = x^2 - \operatorname{tr}(x)x + \frac{1}{2}(\operatorname{tr}^2(x) - \operatorname{tr}(x^2)) = 0.$$

Now we linearize the identity $c(x) = 0$ and obtain the *mixed Cayley–Hamilton identity*

$$\Psi_2(x_1, x_2) = x_1 x_2 + x_2 x_1 - \operatorname{tr}(x_1)x_2 - \operatorname{tr}(x_2)x_1 + \operatorname{tr}(x_1)\operatorname{tr}(x_2) - \operatorname{tr}(x_1 x_2) = 0.$$

Since the trace is a non-degenerated bilinear form on $M_2(K)$, the vanishing of the polynomial $\Psi_2(x_1, x_2)$ on $M_2(K)$ is equivalent to the vanishing of the *pure Cayley–Hamilton identity*

$$\Phi_2(x_1, x_2, x_3) = \operatorname{tr}(\Psi_2(x_1, x_2)x_3) = 0$$

on all 2×2 matrices, i.e., $\operatorname{tr}(\Psi_2(y_1, y_2)y_3)$ is equal to 0 in the trace ring T_{2d}, $d \geq 3$. Direct calculations show that

$$
\begin{aligned}
0 = \Phi_2(x_1, x_2, x_3) &= \operatorname{tr}(\Psi_2(x_1, x_2)x_3) \\
&= \operatorname{tr}(x_1 x_2 x_3) + \operatorname{tr}(x_2 x_1 x_3) - \operatorname{tr}(x_1)\operatorname{tr}(x_2 x_3) \\
&\quad -\operatorname{tr}(x_2)\operatorname{tr}(x_1 x_3) + \operatorname{tr}(x_1)\operatorname{tr}(x_2)\operatorname{tr}(x_3) - \operatorname{tr}(x_1 x_2)\operatorname{tr}(x_3) \\
&= \operatorname{tr}_{(123)}(x_1, x_2, x_3) + \operatorname{tr}_{(213)}(x_1, x_2, x_3) - \operatorname{tr}_{(23)(1)}(x_1, x_2, x_3) \\
&\quad -\operatorname{tr}_{(13)(2)}(x_1, x_2, x_3) + \operatorname{tr}_{(1)(2)(3)}(x_1, x_2, x_3) - \operatorname{tr}_{(12)(3)}(x_1, x_2, x_3) \\
&= \sum_{\sigma \in S_3} (\operatorname{sign} \sigma)\operatorname{tr}_\sigma(x_1, x_2, x_3).
\end{aligned}
$$

Similar polynomials $\Psi_n(x_1, \ldots, x_n)$ and

$$\Phi_n(x_1, \ldots, x_n, x_{n+1}) = \operatorname{tr}\left(\Psi_n(x_1, \ldots, x_n)x_{n+1}\right)$$

are obtained from the Cayley–Hamilton theorem for any $n \geq 2$ and are called, respectively, the *n-th mixed* and *pure formal Cayley–Hamilton polynomial*. The pure polynomial Φ_n satisfies the equation

$$\Phi_n(x_1, \ldots, x_{n+1}) = \sum_{\sigma \in S_{n+1}} (\operatorname{sign} \sigma)\operatorname{tr}_\sigma(x_1, \ldots, x_{n+1}).$$

The polynomial Ψ_n is an element of the *mixed free trace algebra* $K\langle T(X), X \rangle$. This algebra is generated by the countable set X and all formal traces $\operatorname{tr}(x_{i_1} \cdots x_{i_m})$. We assume that the formal traces are in the centre of $K\langle T(X), X \rangle$ and are invariant under cyclic permutations,

$$\operatorname{tr}(x_{i_1} x_{i_2} \cdots x_{i_m}) = \operatorname{tr}(x_{i_2} \cdots x_{i_m} x_{i_1}).$$

Remark 5.2.5. The trace identities are naturally related with the ordinary polynomial identities of matrices. Since the trace is a nondegenerate bilinear form, the multilinear polynomial $f(x_1, \ldots, x_m) \in K\langle X \rangle$ is an identity of $M_n(K)$ if and only if $\operatorname{tr}(f(x_1, \ldots, x_m)x_{m+1}) = 0$ is a trace identity. Each summand of

$$\operatorname{tr}(f(x_1, \ldots, x_m)x_{m+1}) = \sum_{\pi \in S_{m+1}} \alpha_\pi \operatorname{tr}(x_{\pi(1)} \cdots x_{\pi(m)} x_{m+1})$$

contains only one trace. Hence $\operatorname{tr}(f(x_1, \ldots, x_m)x_{m+1})$ is a linear combination of $\operatorname{tr}_\sigma(x_1, \ldots, x_m, x_{m+1})$ where all $\sigma = (\pi(1), \ldots, \pi(m), m+1)$ are cycles of length $m+1$ in S_{m+1}. Therefore the set of ordinary multilinear polynomial identities of degree m for $M_n(K)$ can be identified with the intersection of the ideal of KS_{m+1} generated by the fundamental trace identity with the subspace of KS_{m+1} spanned by all cycles of length $m+1$. The ideal of KS_{m+1} generated by the fundamental trace identity has several natural combinatorial descriptions and is easier to study than the set of the ordinary multilinear identities. Hence the above relation can be used (and has already been used) for estimates of the growth of the polynomial identities of $M_n(K)$.

Considering the defining relations of the algebra C_{nd} for a fixed d, it is known that the transcendence degree of C_{nd} is

$$\text{transc. deg. } (C_{nd}) = (d-1)n^2 + 1,$$

i.e., the algebra C_{nd} contains $(d-1)n^2 + 1$ algebraically independent elements and every $(d-1)n^2 + 2$ elements satisfy a non-trivial algebraic equation.

We shall discuss the defining relations of C_{2d} in the next section. For $n \geq 3$, a first approximation to the concrete description of the defining relations (and an important problem of independent interest) is the computation of the Hilbert series of the algebra of matrix invariants. The concrete form of the Hilbert series of C_{32} and C_{42} as a bigraded vector space was given by Teranishi [T1, T2]. We give it for two 3×3 matrices only (the commuting variables t and u count, respectively, the degrees of x and y).

$$H = H(C_{32}, t, u)$$
$$= \frac{1 + t^3 u^3}{(1 - t^2 u^2)(1 - t^2)(1 - tu)(1 - u^2)(1 - t^3)(1 - t^2 u)(1 - tu^2)(1 - u^3)}.$$

On the other hand, the first ten of the generators of C_{32} in the list of Teranishi (5.1) are algebraically independent, i.e., generate the polynomial algebra $K[Z_{10}] = K[z_1, \ldots, z_{10}]$. The eleventh generator $\operatorname{tr}(x^2 y^2 xy)$ of C_{32}, together with 1, spans C_{32} as a free $K[Z_{10}]$-module of rank 2. Hence $\operatorname{tr}(x^2 y^2 xy)$ satisfies a quadratic relation with coefficients depending on the other ten traces. This relation was recently given explicitly, by complicated formulas, by Nakamoto [Nk].

Now we survey some results devoted to the Hilbert series of C_{nd}. The commutative algebra C_{nd} is generated by a finite system of multihomogeneous elements. A general result of commutative algebra gives that the Hilbert series of C_{nd} is a rational function of $t_1, \ldots, t_d$. The Weyl analogue for compact groups of the Molien formula [We1] gives an expression for the Hilbert series of C_{nd} as a multiple integral. Recall that the complex unitary group $U(n, \mathbb{C})$ consists of all invertible $n \times n$

matrices g with complex entries such that $g^{-1} = \bar{g}^t$, where $\bar{g}^t$ is the transposed matrix of the complex conjugated of g.

Theorem 5.2.6. (see, e.g., [F3], p. 204) *Let $U(n, \mathbb{C})$ be the complex unitary group with normalized Haar measure μ. Then the Hilbert series of C_{nd} is given by*

$$H(C_{nd}, t_1, \ldots, t_d) = \int_{U(n,\mathbb{C})} \left(\prod_{i=1}^{d} \frac{1}{\det(1 - t\varphi^g)} \right) d\mu(g),$$

where for each $g \in U(n, \mathbb{C})$, $\varphi^g \in U(n^2, \mathbb{C})$ is a matrix giving the action of g by conjugation on $M_n(\mathbb{C})$.

The evaluation of the multiple integrals from Theorem 5.2.6 is difficult. For example, the explicit form of the Hilbert series of C_{32} and C_{42} established by Teranishi was obtained by application of the Cauchy integral formula to the integral in Theorem 5.2.6.

Since C_{nd} has additional nice algebraic properties, its Hilbert series should be nicely looking. Van den Bergh [VB2] used some ideas of Stanley [St2] to reduce the determination of the Hilbert series of C_{nd} to a problem about flows in a certain graph and obtained an important consequence for the denominators of the rational functions in the explicit form of the series. He also evaluated some more $H(C_{nd}, t_1, \ldots, t_d)$ for small n and d.

Theorem 5.2.7. (Van den Bergh [VB2]) *The Hilbert series of C_{nd} can be expressed as a rational function whose denominator is a product of terms $(1 - u)$ where u is a monomial in $t_1, \ldots, t_d$ of degree $\leq n$.*

Another confirmation for the nice properties of C_{nd} is that its Hilbert series satisfies a functional equation.

Theorem 5.2.8. (Formanek [F5], Teranishi [T1], see also Le Bruyn [LB1] for the case $n = 2$) *Let $H(t_1, \ldots, t_d)$ be the Hilbert series of C_{nd}, where $d \geq 2$ for $n \geq 3$ and $d > 2$ for $n = 2$. Then $H(t_1, \ldots, t_d)$ satisfies the functional equation*

$$H(t_1^{-1}, \ldots, t_d^{-1}) = (-1)^{\mathrm{Kdim}} (t_1 \cdots t_d)^{n^2} H(t_1, \ldots, t_d),$$

where $\mathrm{Kdim} = \text{transc. deg.} (C_{nd}) = (d - 1)n^2 + 1$ is the Krull dimension of C_{nd}.

The proofs of Theorem 5.2.8 given by Formanek and Teranishi are quite different and use, respectively, representation theory of general linear groups and the Weyl integral formula. Later, Van den Bergh paid attention that the proof can be considerably simplified using results of Stanley on Hilbert series of Cohen–Macaulay algebras.

Another description of the Hilbert series of C_{nd} in the language of representation theory of GL_d is given by Formanek [F3]; see also [F2] for references how to translate the results of Razmyslov–Procesi [Ra3, Pr3] in order to obtain another expression of the Hilbert series of C_{nd}.

   Most of our attention in this section was on the algebra C_n which is the algebra of invariants of $GL_n(K)$ acting on the polynomial algebra Ω_n. The mixed trace algebra T_n generated by the generic matrices $y_1, y_2, \ldots$ and the elements of C_n also has a similar description.

   Recall that if G is a group and V and W are G-modules then G acts *diagonally* on the tensor product $V \otimes W$ by

$$g(v \otimes w) = g(v) \otimes g(w), \quad g \in G, v \in V, w \in W.$$

In particular, if G acts on an algebra R and on the $n \times n$ matrix algebra $M_n(K)$ we define the diagonal action on the matrix algebra $M_n(R) \cong R \otimes_K M_n(K)$ with entries from R via

$$g\left(\sum_{p=1}^{n}\sum_{q=1}^{n} r_{pj}e_{pq}\right) = \sum_{p=1}^{k}\sum_{q=1}^{k} g(r_{pq})g(e_{pq}), \quad g \in G, r_{pq} \in R.$$

Theorem 5.2.9. (Procesi [Pr3], Section 2) *Let $GL_n(K)$ act on Ω_n as described above and on $M_n(K)$ by $g(a) = g^{-1}ag$, where $a \in M_n(K)$ and $g \in GL_n(K)$. (Note that g^{-1} and g are reversed in comparison to the action of $GL_n(K)$ on Ω induced by the action on the generic matrices $y_i = (y_{pq}^{(i)})$.) This induces a diagonal action of $GL_n(K)$ on $M_n(\Omega_n) \cong \Omega_n \otimes_K M_n(K)$ and the mixed trace algebra T_n is the fixed algebra of this action.*

   In virtue of this result the mixed trace algebra is called also the *algebra of matrix concomitants*. A lot of the results on C_n and C_{nd} are true also for the noncommutative algebras T_n and T_{nd}. For example, the Hilbert series of T_{nd} satisfies the same functional equation given in Theorem 5.2.8, see Formanek [F5], Teranishi [T3], and Le Bruyn [LB1] for $n = 2$.

Theorem 5.2.10. (Van den Bergh [VB1]) *The algebra T_{nd} is a Cohen–Macaulay module over C_{nd}.*

   Recall that the Noether normalization theorem gives that C_{nd} contains a homogeneous set of algebraically independent elements $\{a_1, \ldots, a_k\}$, where $k = (d-1)n^2 + 1$ is the transcendence degree of the quotient field of C_{nd}, such that C_{nd} is integral over the polynomial algebra $K[a_1, \ldots, a_k]$. Such a set $\{a_1, \ldots, a_k\}$ is called a *homogeneous system of parameters* for C_{nd}. By a result of Stanley [St1], a graded C_{nd}-module is Cohen–Macaulay if and only if it is a free module with respect to some homogeneous system of parameters $\{a_1, \ldots, a_k\}$ of C_{nd}. Theorems 5.2.2 and 5.2.10 give rise to the following problem.

Problem 5.2.11. *Find bases of C_{nd} and T_{nd} as free modules over $K[a_1, \ldots, a_k]$ for some homogeneous system of parameters $\{a_1, \ldots, a_k\}$ for C_{nd}.*

5.3   Invariants of 2×2 Matrices

First we summarize some results on the algebra C_{2d}, $d \geq 2$.

Theorem 5.3.1. (i) (First Fundamental Theorem of 2×2 Matrix Invariants, Sibirskii [Si]) *The polynomials*

$$\operatorname{tr}(y_i),\ i = 1,\ldots,d, \quad \operatorname{tr}(y_i y_j),\ 1 \le i \le j \le d,$$
$$\operatorname{tr}(y_i y_j y_k),\ 1 \le i < j < k \le d,$$

form a minimal system of generators of C_{2d}.

(ii) (Second Fundamental Theorem of Invariants of Two and Three 2×2 Matrices, Sibirskii [Si], Formanek [F3]) *Let*

$$a_i = \operatorname{tr}(y_i), b_i = \det(y_i), c_i = \operatorname{tr}(y_j y_k), d = \operatorname{tr}(y_1 y_2 y_3),$$

$i,j,k = 1,2,3,\ i \ne j,k,\ j < k$. *The algebra of invariants C_{22} of two generic 2×2 matrices is isomorphic to the polynomial algebra $K[a_1, a_2, b_1, b_2, c_3]$. The algebra of invariants of three generic matrices C_{23} is generated by the commuting elements $a_i, b_i, c_i, d,\ i = 1,2,3$, modulo the principal ideal generated by the relation*

$$d^2 - (a_1 c_1 + a_2 c_2 + a_3 c_3 - a_1 a_2 a_3)d$$
$$+ b_1 c_1^2 + b_2 c_2^2 + b_3 c_3^2 - a_1 a_2 b_3 c_3 - a_3 a_1 b_2 c_2 - a_2 a_3 b_1 c_1$$
$$+ a_1^2 b_2 b_3 + a_2^2 b_3 b_1 + a_3^2 b_1 b_2 - 4 b_1 b_2 b_3 + c_1 c_2 c_3 = 0.$$

Alternatively, C_{23} is a free module of rank 2 with basis $\{1, d\}$ over the polynomial algebra $K[a_i, b_i, c_i \mid i = 1,2,3]$.

Concrete computer calculations using programs for computing Gröbner bases were performed for small d by Aslaksen, Tan and Zhu [ATZ]. They also found minimal systems of generators and some defining relations for the invariants of the classical subgroups of $GL_2(K)$.

Since the transcendence degree of C_{2d} is

$$\text{transc. deg. } (C_{2d}) = (d-1)2^2 + 1 = 4d - 3,$$

the algebra C_{2d} contains $4d - 3$ algebraically independent elements and every $4d - 2$ elements satisfy a non-trivial algebraic equation. The precise result sounds as follows.

Theorem 5.3.2. (see [ATZ], Theorem 6) *The polynomials*

$$\operatorname{tr}(x_i), \operatorname{tr}(x_i^2), \operatorname{tr}(x_1 x_j), \operatorname{tr}(x_2 x_k), \quad i = 1,\ldots,d,\ j = 2,\ldots,d,\ k = 3,\ldots,d,$$

form a maximal set of algebraically independent elements of C_{2d}.

Below we give the complete description of the algebra of invariants C_{2d}. For this purpose, we use invariant theory of the orthogonal group $SO_3(\mathbb{C})$, as in the book by Le Bruyn [LB2]. We follow the exposition of the paper by the author [Dr11]. Until the end of the section we replace K with $\mathbb{C}$. It is easy to see that the obtained finite results hold for any field K of characteristic 0.

The action of $g \in GL_2(\mathbb{C})$ on Ω_{2d} does not depend on the determinant of g and one may replace the action of GL_2 with that of the special linear group $SL_2 = SL_2(\mathbb{C})$ and even of $PSL_2 = PSL_2(\mathbb{C})$. The group PSL_2 is isomorphic to the special orthogonal group $SO_3 = SO_3(\mathbb{C})$ and this allows to involve to the description of the algebra of invariants of $\Omega_{2d}^{GL_2(\mathbb{C})}$ classical results on invariant theory of orthogonal groups. First we give the minimal background on invariant theory of orthogonal groups which we need.

Let us recall the definition of the orthogonal group. We fix an n-dimensional complex vector space V_n with a nondegenerate symmetric bilinear form. Without loss of generality we may assume that V_n has a basis $\{e_1, \ldots, e_n\}$ such that the form coincides with the scalar product

$$\langle a, b \rangle = \sum_{i=1}^{n} \xi_i \eta_i, \quad a = \sum_{i=1}^{n} \xi_i e_i, \quad b = \sum_{i=1}^{n} \eta_i e_i, \quad \xi_i, \eta_i \in \mathbb{C}.$$

Then the orthogonal group $O_n = O_n(\mathbb{C}) = O(V_n)$ coincides with the set of matrices $g \in GL_n = GL_n(\mathbb{C})$ preserving the form (i.e., $\langle g(a), g(b) \rangle = \langle a, b \rangle$ for all a, b in V_n), and $SO_n = SO_n(\mathbb{C})$ is the subgroup of matrices in O_n with determinant 1.

For invariant theory of SO_n we fix an nd-dimensional complex vector space U_{nd} with basis

$$\{u_{ij} \mid i = 1, \ldots, n, \quad j = 1, \ldots, d\},$$

the polynomial algebra $\mathbb{C}[U_{nd}]$, and consider the "generic" vectors

$$u_j = (u_{1j}, \ldots, u_{nj}), \quad j = 1, \ldots, d,$$

with scalar product

$$\langle u_i, u_j \rangle = u_{1i} u_{1j} + \cdots + u_{ni} u_{nj}, \quad 1 \leq i, j \leq d.$$

The group $GL_n(\mathbb{C})$ acts naturally on each n-tuple

$$g(u_j) = v_j = (v_{1j}, \ldots, v_{nj}), \quad g \in GL_n(\mathbb{C}),$$

and this induces the action

$$g(u_{ij}) = v_{ij}.$$

We define the polynomials in the coordinates u_{ij}

$$\Delta_n(u_{j_1}, \ldots, u_{j_n}) = \det(u_{j_1}, \ldots, u_{j_n}) = \begin{vmatrix} u_{1j_1} & u_{1j_2} & \cdots & u_{1j_n} \\ u_{2j_1} & u_{2j_2} & \cdots & u_{2j_n} \\ \vdots & \vdots & \cdots & \vdots \\ u_{nj_1} & u_{nj_2} & \cdots & u_{nj_n} \end{vmatrix},$$

$$\Gamma_k(u_{i_1}, \ldots, u_{i_k} \mid u_{j_1}, \ldots, u_{j_k}) = \det(\langle u_{i_p}, u_{j_q} \rangle) = \begin{vmatrix} \langle u_{i_1}, u_{j_1} \rangle & \cdots & \langle u_{i_1}, u_{j_k} \rangle \\ \vdots & \cdots & \vdots \\ \langle u_{i_k}, u_{j_1} \rangle & \cdots & \langle u_{i_k}, u_{j_k} \rangle \end{vmatrix}.$$

We assume that Δ_n and Γ_k are equal to 0 and do not participate in the corresponding systems of generators when $n > d$ or $k > d$. The polynomial $f = f(u_1, \ldots, u_d) \in \mathbb{C}[U_{nd}]$ is called an *absolute* (or *even*) *orthogonal invariant* if $g(f) = f$ for all $g \in O_n$ and an *odd invariant* if $g(f) = f$ for $g \in SO_n$ and $g(f) = -f$ for $g \in O_n - SO_n$. The description of the generators and the defining relations of O_n and SO_n is the following (see, e.g., Weyl [We2]).

Theorem 5.3.3. (i) *The algebra of absolute orthogonal invariants* $\mathbb{C}[U_{nd}]^{O_n}$ *is generated by the scalar products* $\langle u_i, u_j \rangle$, $1 \le i \le j \le d$.

(ii) *The algebra of invariants* $\mathbb{C}[U_{nd}]^{SO_n}$ *is a sum of the absolute orthogonal invariants* $\mathbb{C}[U_{nd}]^{O_n}$ *and of the odd invariants* $\Delta(u_{i_1}, \ldots, u_{i_n})\mathbb{C}[U_{nd}]^{O_n}$, $1 \le i_1 < \cdots < i_n \le d$.

Theorem 5.3.4. (i) *The defining relations of* $\mathbb{C}[U_{nd}]^{O_n}$ *are*

$$\Gamma_{n+1}(u_{i_0}, u_{i_1}, \ldots, u_{i_n} \mid u_{j_0}, u_{j_1}, \ldots, u_{j_n}) = 0,$$

$$1 \le i_0 < i_1 < \cdots < i_n \le d, \ 1 \le j_0 < j_1 < \cdots < j_n \le d.$$

(ii) *The defining relations of* $\mathbb{C}[U_{nd}]^{SO_n}$ *are*

$$\Gamma_{n+1}(u_{i_0}, u_{i_1}, \ldots, u_{i_n} \mid u_{j_0}, u_{j_1}, \ldots, u_{j_n}) = 0,$$

$$1 \le i_0 < i_1 < \cdots < i_n \le d, \ 1 \le j_0 < j_1 < \cdots < j_n \le d,$$

$$\Delta_n(u_{i_1}, \ldots, u_{i_n})\Delta_n(u_{j_1}, \ldots, u_{j_n}) - \Gamma_n(u_{i_1}, \ldots, u_{i_n} \mid u_{j_1}, \ldots, u_{j_n}) = 0,$$

$$1 \le i_1 < \cdots < i_n \le d, \ 1 \le j_1 < \cdots < j_n \le d,$$

$$\sum_{k=0}^{n} (-1)^k \langle u_i, u_{j_k} \rangle \Delta_n(u_{j_0}, \ldots, \hat{u}_{j_k}, \ldots, u_{j_n}) = 0, \ 1 \le j_0 < j_1 < \cdots < j_n \le d,$$

where $\hat{u}_{j_k}$ *means that* u_{j_k} *does not participate in the expression.*

Remark 5.3.5. One can show, see, e.g., [Dr11] that the defining relations of the algebra of invariants $\mathbb{C}[U_{nd}]^{SO_n}$ of the type

$$\Gamma_{n+1}(u_{i_0}, u_{i_1}, \ldots, u_{i_n} \mid u_{j_0}, u_{j_1}, \ldots, u_{j_n}) = 0,$$

are consequences of the relations of the other two types

$$\Delta_n(u_{i_1}, \ldots, u_{i_n})\Delta_n(u_{j_1}, \ldots, u_{j_n}) - \Gamma_n(u_{i_1}, \ldots, u_{i_n} \mid u_{j_1}, \ldots, u_{j_n}) = 0,$$

$$\sum_{k=0}^{n} (-1)^k \langle u_i, u_{j_k} \rangle \Delta_n(u_{j_0}, \ldots, \hat{u}_{j_k}, \ldots, u_{j_n}) = 0.$$

Remark 5.3.6. We fix the vector space Tr_d with basis $\{t_1, \ldots, t_d\}$ and, together with the vector space U_{nd} with basis $\{u_{ij} \mid i = 1, \ldots, n, \ j = 1, \ldots, d\}$ defined above, we consider the polynomial algebra

$$\mathbb{C}[\text{Tr}_d, U_{nd}] = \mathbb{C}[\text{Tr}_d] \otimes \mathbb{C}[U_{nd}]$$

(the tensor products are over $\mathbb{C}$) with the trivial action of SO_n on $\mathbb{C}[\text{Tr}_d]$. The algebra $\mathbb{C}[\text{Tr}_d, U]^{SO_n}$ coincides with $\mathbb{C}[\text{Tr}_d] \otimes \mathbb{C}[U]^{SO_n}$ and is generated by t_i, $i = 1, \ldots, d$, and the generators of $\mathbb{C}[U]^{SO_n}$ described in Theorem 5.3.3. Its defining relations coincide with the defining relations of $\mathbb{C}[U]^{SO_n}$ given in Theorem 5.3.4.

Now we are ready to translate the results of Theorems 5.3.3 and 5.3.4 to the language of matrix invariants and to find systems of generators and defining relations for the invariants of $GL_2(\mathbb{C})$ acting by simultaneous conjugation on d-tuples of 2×2 matrices. It turned out that the traditional generators of the algebras of 2×2 matrix invariants given in Theorem 5.3.1 are not very convenient for our purposes and we shall give another version of the first and second fundamental theorem inspired by the approaches used in the study by Procesi [Pr4] and Le Bruyn [LB2] on generic 2×2 matrices and by Alev and Le Bruyn [ALB] on automorphisms of generic 2×2 matrices.

Let $V_3 = sl_2 = sl_2(\mathbb{C})$ be the three-dimensional vector space of all traceless 2×2 matrices. We consider the nondegenerate symmetric bilinear form

$$\langle u, v \rangle = \mathrm{tr}(uv), \ u, v \in sl_2,$$

and fix the basis in V_3

$$e_1 = \frac{\sqrt{2}}{2}(e_{11} - e_{22}), \quad e_2 = \frac{\sqrt{2}}{2}(e_{12} + e_{21}), \quad e_3 = i\frac{\sqrt{2}}{2}(e_{12} - e_{21}),$$

where $i^2 = -1$ in $\mathbb{C}$. It is easy to see that $\langle e_j, e_k \rangle = \delta_{jk}$, where δ_{jk} is the Kronecker symbol, i.e., the basis is orthogonal and normed with respect to the bilinear form. Let us consider the action of $PSL_2 = PSL_2(\mathbb{C})$ on V_3 by conjugation. Since PSL_2 is simple also as an abstract group, it acts faithfully on V_3 and preserves the bilinear form. Hence, the action of PSL_2 is the same as the action of $O_3 = O(V_3)$ on V_3, and there is a natural embedding $\varphi : PSL_2 \rightarrow O_3 = O(V_3) \subset GL(V_3)$. The following lemma gives a well known fact in the theory of classical groups.

Lemma 5.3.7. *The mapping φ is an isomorphism of PSL_2 and $SO_3 \subset GL(V_3)$.*

We consider the 4-dimensional vector space $V_4 = M_2(\mathbb{C})$ as a direct sum

$$V_4 = M_2(\mathbb{C}) = \mathbb{C}\frac{e_{11} + e_{22}}{2} \oplus sl_2(\mathbb{C}),$$

both subspaces being invariant under the conjugation by $GL_2(\mathbb{C})$. We also fix the basis $\{e_1, e_2, e_3\}$ of sl_2 which is orthogonal and normed with respect to the bilinear form $\langle u, v \rangle = \mathrm{tr}(uv)$. The 4$d$-dimensional vector space has a basis consisting of the entries of the d generic matrices

$$y_j = \begin{pmatrix} y_{11}^{(j)} & y_{12}^{(j)} \\ y_{21}^{(j)} & y_{22}^{(j)} \end{pmatrix}, \quad j = 1, \ldots, d.$$

We obtain that

$$y_j = t_j \frac{e_{11} + e_{22}}{2} + z_j,$$

where $t_j = \mathrm{tr}(y_j)$ and

$$z_j = \begin{pmatrix} z_{11}^{(j)} & z_{12}^{(j)} \\ z_{21}^{(j)} & -z_{11}^{(j)} \end{pmatrix}, \quad j = 1, \ldots, d,$$

are d generic 2×2 traceless matrices with entries related with the entries of y_j by

$$z_{11}^{(j)} = \frac{1}{2} \left(y_{11}^{(j)} - y_{22}^{(j)} \right), \quad z_{12}^{(j)} = y_{12}^{(j)}, \quad z_{21}^{(j)} = y_{21}^{(j)}.$$

The group $GL_2(\mathbb{C})$ acts trivially on $(e_{11} + e_{22})/2$. Hence the polynomial algebra in $4d$ variables where we search for the matrix invariants is $\mathbb{C}[\mathrm{Tr}_d, U_{3d}]$, where

$$\mathrm{Tr}_d = \mathrm{span}\{t_1, \ldots, t_m\}, \quad U_{3d} = \mathrm{span}\{u_{1j}, u_{2j}, u_{3j} \mid j = 1, \ldots, d\},$$

$t_i = \mathrm{tr}(y_i)$, and the generic traceless matrix z_j is of the form

$$y_j = u_{1j}e_1 + u_{2j}e_2 + u_{3j}e_3 = \frac{\sqrt{2}}{2} \left(2z_{11}^{(j)}e_1 + (z_{12}^{(j)} + z_{21}^{(j)})e_2 + i(-z_{12}^{(j)} + z_{21}^{(j)})e_3 \right).$$

We see by direct calculations that

$$s_3(e_1, e_2, e_3) = \sum_{\sigma \in S_3} (\mathrm{sign}\ \sigma)e_{\sigma(1)}e_{\sigma(2)}e_{\sigma(3)} = -\frac{3i}{2}\sqrt{2}(e_{11} + e_{22}),$$

$$\mathrm{tr}(s_3(e_1, e_2, e_3)) = -3i\sqrt{2},$$

$$\Delta_3(z_1, z_2, z_3) = i\frac{\sqrt{2}}{6}\mathrm{tr}(s_3(z_1, z_2, z_3)).$$

The two parts of the following theorem [LB2, Dr11] are the first and the second fundamental theorems of the 2×2 matrix invariants. (The sets of defining relations look uniformly for all $d \geq 2$. Compare the generators in the first part with the usual set of generators $\{\mathrm{tr}(y_i), \mathrm{tr}(y_j y_k), \mathrm{tr}(y_{i_1} y_{i_2} y_{i_3})\}$ and with the set $\{\mathrm{tr}(y_i), \mathrm{tr}(z_j z_k), \mathrm{tr}(z_{i_1} z_{i_2} z_{i_3})\}$ given by Le Bruyn [LB2], Ch. 1, Proof of Theorem 4.2, where in both sets $1 \leq i \leq d$, $1 \leq j \leq k \leq d$, $1 \leq i_1 < i_2 < i_3 \leq d$.)

Theorem 5.3.8. *Let $y_1, \ldots, y_d$ be generic 2×2 matrices and let $z_1, \ldots, z_d$ be generic traceless 2×2 matrices over the complex field $\mathbb{C}$.*

(i) The algebra of $GL_2(\mathbb{C})$-invariants under simultaneous conjugation of d-tuples of 2×2 matrices is generated by

$$\mathrm{tr}(y_i), \mathrm{tr}(z_j z_k), \mathrm{tr}(s_3(z_{j_1}, z_{j_2}, z_{j_3})),$$

where $1 \leq i \leq d$, $1 \leq j \leq k \leq d$, $1 \leq j_1 < j_2 < j_3 \leq d$, and

$$s_3(z_1, z_2, z_3) = \sum_{\sigma \in S_3} (\mathrm{sign}\ \sigma)z_{\sigma(1)}z_{\sigma(2)}z_{\sigma(3)}$$

is the standard polynomial of degree 3.

(ii) *The defining relations of the algebra of $GL_2(\mathbb{C})$-invariants with respect to the above generators are*

$$\operatorname{tr}(s_3(z_{i_1}, z_{i_2}, z_{i_3}))\operatorname{tr}(s_3(z_{j_1}, z_{j_2}, z_{j_3})) + 18 \begin{vmatrix} \operatorname{tr}(z_{i_1}z_{j_1}) & \operatorname{tr}(z_{i_1}z_{j_2}) & \operatorname{tr}(z_{i_1}z_{j_3}) \\ \operatorname{tr}(z_{i_2}z_{j_1}) & \operatorname{tr}(z_{i_2}z_{j_2}) & \operatorname{tr}(z_{i_2}z_{j_3}) \\ \operatorname{tr}(z_{i_3}z_{j_1}) & \operatorname{tr}(z_{i_3}z_{j_2}) & \operatorname{tr}(z_{i_3}z_{j_3}) \end{vmatrix} = 0,$$

$$\sum_{k=0}^{3}(-1)^k\operatorname{tr}(z_iz_{p_k})\operatorname{tr}(s_3(z_{p_0}, \ldots, \hat{z}_{p_k}, \ldots, z_{p_3})) = 0,$$

$$1 \le i_1 < i_2 < i_3 \le d, \quad 1 \le j_1 < j_2 < j_3 \le d,$$
$$1 \le i \le d, \quad 1 \le p_0 < p_1 < p_2 < p_3 \le d.$$

Proof. By Lemma 5.3.7, the action of $GL_2(\mathbb{C})$ on the vector space $V_3 = sl_2$ coincides with the action of $SO_3(\mathbb{C})$ on the three-dimensional vector space equipped with a symmetric nondegenerate bilinear form. Also, $GL_2(\mathbb{C})$ acts on the scalar matrices trivially. Hence, we may apply Remark 5.3.6. Now, it is sufficient to restate Theorems 5.3.3 (ii) and 5.3.4 (ii) as well as Remark 5.3.5 in our language. Part (i) of the theorem follows immediately from the fact that $\operatorname{tr}(z_jz_k)$ is equal to the scalar product $\langle z_j, z_k \rangle$ and that $\Delta_3(z_{i_1}, z_{i_2}, z_{i_3})$ and $\operatorname{tr}(s_3(z_{i_1}, z_{i_2}, z_{i_3}))$ are proportional. For part (ii) it is sufficient to replace the expression of $\Delta_3(z_{i_1}, z_{i_2}, z_{i_3})$ by $\operatorname{tr}(s_3(z_{i_1}, z_{i_2}, z_{i_3}))$. $\square$

Remark 5.3.9. Aslaksen, Tan and Zhu [ATZ] found nontrivial relations of degree 8 between the 2×2 matrix invariants which involve 4 generic matrices. This is not a contradiction with Theorem 5.3.8 (ii) because in [ATZ] the authors look for the Gröbner basis of the ideal generated by the defining relations. Very often the Gröbner basis of an ideal contains more elements than the minimal set of generators. In our case Remark 5.3.5 gives that all relations of degree 8 are consequences of the relations of lower degree.

In a similar way one can handle the generators and the defining relations of the algebra of matrix concomitants. As above, $y_1, \ldots, y_d$ are d generic 2×2 matrices, $z_1, \ldots, z_d$ are d generic traceless 2×2 matrices and K is an arbitrary field of characteristic 0. We shall state the theorem, with some ideas of the proof only. It shows that the algebra of 2×2 generic traceless matrices has a uniform set of defining relations for any $d \ge 2$.

Theorem 5.3.10. (i) *The mixed 2×2 trace algebra T_{2d} is isomorphic to the tensor product*

$$K[\operatorname{tr}(y_1), \ldots, \operatorname{tr}(y_d)] \otimes W_d,$$

where W_d is the associative algebra generated by the generic traceless matrices $z_1, \ldots, z_d$.

(ii) (Razmyslov [Ra1]) *The algebra W_d is isomorphic to the factor algebra of $K\langle x_1, \ldots, x_d \rangle$ modulo the ideal generated by all $[u^2, v]$ where u, v run on the Lie subalgebra of $K\langle x_1, \ldots, x_d \rangle$ generated by $x_1, \ldots, x_d$.*

(iii) (Drensky, Koshlukov [DKo]) *The algebra W_d has as defining relations the set of all polynomials $[x_i x_j + x_j x_i, x_k]$, $i, j, k = 1, \ldots, d$, and the standard polynomials $s_4(x_{i_1}, x_{i_2}, x_{i_3}, x_{i_4})$, $1 \leq i_1 < i_2 < i_3 < i_4 \leq d$, the second kind of relations appears for $d \geq 4$ only.*

Proof. The statement of (i) uses the arguments that the generic matrix y_i is a sum of the scalar matrix $\mathrm{tr}(y_i)e/2$ and the traceless generic matrix z_i. Since $\mathrm{tr}(y_i)e$ belongs to T_{2d}, the same holds for z_i and the algebra T_{2d} contains the algebras generated by $\mathrm{tr}(y_1), \ldots, \mathrm{tr}(y_d)$ and by $z_1, \ldots, z_d$. Clearly, the commutative algebra generated by the traces has no relations and is isomorphic to $K[\mathrm{tr}(y_1), \ldots, \mathrm{tr}(y_d)]$. Similarly one proves that $K[\mathrm{tr}(y_1), \ldots, \mathrm{tr}(y_d)]$ and W_d participate in T_{2d} as a tensor product. In order to show the coincidence of the tensor product with the whole T_{2d} it is sufficient to pay attention to the fact that z_i^2 and $z_i z_j + z_j z_i$ commute with the elements of W_d (the weak polynomial identity $[x_1^2, x_2] = 0$ for $M_2(K)$!). One can see directly that they are scalar matrices. Hence the generators $\mathrm{tr}(z_i^2)$ and $\mathrm{tr}(z_i z_j)$ of C_{2d} belong to the algebra W_d. The other generators of C_{2d}, namely $\mathrm{tr}(s_3(z_i, z_j, z_k))$ also are proportional to the scalar matrices $s_3(z_i, z_j, z_k)$ and belong to W_d.

(ii) This part of the theorem follows from the theorem of Razmyslov [Ra1] stating that all weak polynomial identities for $M_2(K)$ are consequences of the weak identity $[x_1^2, x_2] = 0$.

(iii) The original proof given in [DKo] was based on combinatorial methods and did not use any invariant theory. Later Koshlukov [Ko] gave another proof which holds also over infinite fields of positive characteristic, where he did calculations based on some classical results on representation theory of GL_d in the spirit of Doubilet, Rota and Stein [DRS] and De Concini and Procesi [DCP].   □

Remark 5.3.11. Some of the results of the theory of matrix invariants over a field of characteristic 0 hold also over an infinite field of positive characteristic. Nevertheless there are many differences between the two cases. See the recent paper [DoKZ] by Domokos, Kuzmin and Zubkov, which focuses on the exact frontiers for validity of the results of characteristic 0 in the case of positive characteristic. In particular, the authors discover a new phenomenon: If $0 < \mathrm{char}\, K \leq n$, then the degree of the generators of R_{nd} goes to infinity with d. The authors study in more detail the case of 2×2 matrices. They show that R_{2d} has a system of generators consisting of $\det(y_i)$ and $\mathrm{tr}(y_{i_1} \cdots y_{i_s})$, where $y_1, \ldots, y_d$ are 2×2 generic matrices and $1 \leq i \leq d$, $1 \leq i_1 < \cdots < i_s \leq d$. If $\mathrm{char}\, K > 2$, then R_{2d} is generated by the determinant and the traces for $s \leq 3$. If $\mathrm{char}\, K = 2$, then one needs the traces of all products for $s \leq d$.

Chapter 6

The Nagata–Higman Theorem

6.1 The Nagata–Higman Theorem and its Applications

The Nagata–Higman theorem for the nilpotency of nil algebras of bounded index was proved in 1953 by Nagata [Na1] over a field of characteristic 0 and then in 1956 by Higman [Hg] in the general setup. Much later it was discovered that this theorem was first established in 1943 by Dubnov and Ivanov [DI] but their paper was overlooked by the mathematical community. The theorem has many applications to the theory of PI-algebras as well as to invariant theory and structure theory of rings. Since we consider nil and nilpotent algebras which are nonunitary, in this chapter we shall work with the free nonunitary algebra $K^+\langle X \rangle$.

Example 6.1.1. The polynomial identity $x^2 = 0$ implies the identity $x_1 x_2 x_3 = 0$. For the proof, we linearize $x^2 = 0$ and obtain

$$e_2(x_1, x_2) = x_1 x_2 + x_2 x_1 = 0.$$

The identity $e_2(x_1 x_2, x_3) = (x_1 x_2) x_3 + x_3 (x_1 x_2) = 0$ gives

$$
\begin{aligned}
x_1 x_2 x_3 &= (x_1 x_2) x_3 = -x_3 (x_1 x_2) = -(x_3 x_1) x_2 \\
&= x_2 (x_3 x_1) = (x_2 x_3) x_1 = -x_1 (x_2 x_3) = -x_1 x_2 x_3,
\end{aligned}
$$

Hence $2 x_1 x_2 x_3 = 0$ and $x_1 x_2 x_3 = 0$.

Since

$$e_2(x_1, x_2) = (x_1 + x_2)^2 - x_1^2 - x_2^2,$$

we obtain that $e_2(x_1, x_2)$ is a linear combination of squares. Direct calculations show that

$$x_1 x_2 x_3 = \frac{1}{2} \left(e_2(x_1 x_2, x_3) + e_2(x_2 x_3, x_1) - e_2(x_3 x_1, x_2) \right)$$

and this gives that $x_1 x_2 x_3$ is also a linear combination of squares.

Theorem 6.1.2. (Nagata–Higman [Na1, Hg], also Dubnov–Ivanov [DI])

(i) *Let n be a positive integer. There exists an integer $d = d(n)$ depending on n only such that the polynomial identity of nilpotency $x_1 \cdots x_d = 0$ is a consequence of the polynomial identity $x^n = 0$.*

(ii) *The T-ideal generated by x^n in the free nonunitary algebra $K^+\langle X \rangle$ coincides with the vector space spanned by all n-th powers. In particular, for $m \geq d = d(n)$ the monomial $x_1 \cdots x_m$ has the form*

$$x_1 \cdots x_m = \sum \alpha_u u^n$$

for some $\alpha_u \in K$ and $u \in K^+\langle X \rangle$.

Proof. (i) We shall work modulo the identity $x^n = 0$ (i.e., in the relatively free algebra $F^+(\mathfrak{N})$ of the variety of nonunitary algebras $\mathfrak{N}$ defined by the polynomial identity $x^n = 0$). The partial linearization of $x^n = 0$ is

$$f(x, y) = x^{n-1}y + x^{n-2}yx + \cdots + xyx^{n-2} + yx^{n-1} = 0.$$

Hence

$$f(x, yz^j)z^{n-j-1} = x^{n-1}yz^{n-1} + x^{n-2}yz^j xz^{n-j-1}$$
$$+ \cdots + xyz^j x^{n-2}z^{n-j-1} + yz^j x^{n-1}z^{n-j-1} = 0,$$

$$\sum_{j=0}^{n-1} f(x, yz^j)z^{n-j-1} = nx^{n-1}yz^{n-1} + \sum_{i=0}^{n-2} x^i yf(z, x^{n-i-1}) = 0.$$

Since $f = 0$, we obtain

$$x^{n-1}yz^{n-1} = 0.$$

By induction, the identity $x^{n-1} = 0$ implies nilpotency. Hence, for some d depending on $n - 1$,

$$x_1 \cdots x_d = \sum_i a_i b_i^{n-1}c_i, \quad x_{d+2} \cdots x_{2d+1} = \sum_j u_j v_j^{n-1}w_j,$$

where b_i, v_j are polynomials (without constant terms) and a_i, c_i, u_j, w_j are constants in K or polynomials. Therefore

$$x_1 \cdots x_d x_{d+1}x_{d+2} \cdots x_{2d+1} = \sum_{i,j} a_i (b_i^{n-1}(c_i x_{d+1}u_j)v_j^{n-1})w_j = 0$$

and this shows that the identity $x_1 \cdots x_{2d+1} = 0$ is a consequence of $x^n = 0$.

(ii) Every element of the T-ideal generated by x^n is a linear combination of elements $w^n, uw^n, w^n v, uw^n v$ for some polynomials u, v, w (without constant terms) in the free algebra. Hence it is sufficient to show that the elements yx^n, $x^n z$ and $yx^n z$ are linear combinations of some $u_1^n, \ldots, u_k^n$. The partial linearizations of any polynomial identity $f(x_1, \ldots, x_m)$ are obtained by Vandermonde arguments and, therefore, are linear combinations of $f(u_1, \ldots, u_m)$, $u_i \in K^+\langle X \rangle$. Hence the partial linearization of x^n

$$f_n(x, y) = x^{n-1}y + x^{n-2}yx + \cdots + xyx^{n-2} + yx^{n-1}$$

is a linear combination of some u_i^n, $u_i \in K^+\langle X\rangle$. We consider $f_n(x + x^2, y)$, take the homogeneous component of degree $n+1$ (Vandermonde arguments again) and obtain

$$(x^n y + y x^n) + (n - 2)(x^n y + x^{n-1} y x + \cdots + x y x^{n-1} + y x^n)$$
$$= (x^n y + y x^n) + (n - 2) f_{n+1}(x, y).$$

Since nx^{n+1} is the homogeneous component of degree $n + 1$ of $(x + x^2)^n$, we obtain that $f_{n+1}(x, y)$ is also a linear combination of some u_i^n. Finally, using that $\mathrm{ad}\, y : u \to [u, y]$ is a derivation (or with direct calculations), we see that

$$x^n y - y x^n = [x^n, y] = x^n \mathrm{ad}\, y = f_n(x, [x, y]),$$

i.e., $x^n y + y x^n$ and $x^n y - y x^n$ are both linear combinations of u_i^n and the same holds for $y x^n$ (similarly for $x^n z$). Now

$$y x^n z = (y x^n) z = \sum \alpha_i u_i^n z = \sum \alpha_i \sum \beta_{ij} v_{ij}^n. \qquad \square$$

Corollary 6.1.3. *As a vector space, the T-ideal of $K^+\langle X\rangle$ generated by the polynomial x^n is spanned by*

$$e_n(u_1, \ldots, u_n) = \sum_{\sigma \in S_n} u_{\sigma(1)} \cdots u_{\sigma(n)}, \qquad (6.1)$$

where $u_1, \ldots, u_n$ are monomials of positive degree in the set of free variables X.

Proof. The identity $x^n = 0$ is equivalent to its complete linearization

$$e_n(x_1, \ldots, x_n) = \sum_{\sigma \in S_n} x_{\sigma(1)} \cdots x_{\sigma(n)}$$

and $e_n(x, \ldots, x) = n! x^n$. By Theorem 6.1.2 (ii), the T-ideal $(x^n)^T$ generated by x^n is spanned by u^n, $u \in K^+\langle X\rangle$. Since x^n is a value of e_n, we obtain that $(x^n)^T$ is spanned by $e_n(u_1, \ldots, u_n)$, $u_1, \ldots, u_n \in K^+\langle X\rangle$. The polynomial $e_n(x_1, \ldots, x_n)$ is multilinear and hence we may assume that $u_1, \ldots, u_n$ are monomials. $\qquad \square$

Remark 6.1.4. Our proof gives an upper bound for the class of nilpotency $d(n)$ in the Nagata–Higman theorem. Starting with the obvious $d(1) = 1$ we use the estimate $d(n) \leq 2d(n-1) + 1$ and obtain that $d(n) \leq 2^n - 1$. The proof works also when char $K = p > n$. The condition char $K = p > n$ is essential. It is easy to construct an algebra (even commutative) which satisfies the identity $x^p = 0$ and is not nilpotent.

Below we give some relations of the Nagata–Higman theorem to polynomial identities and invariants of matrices. Recall that $K\langle T(X), X\rangle$ is the mixed free trace algebra, T_n is the mixed trace generic matrix algebra, i.e., the algebra generated by the generic $n \times n$ matrices $Y = \{y_1, y_2, \ldots\}$ and the algebra C_n generated by all traces $\mathrm{tr}(y_{i_1} \cdots y_{i_m})$. We denote by $K^+\langle T(X), X\rangle$ and T_n^+ the nonunitary versions of these algebras. Finally, $\Psi_n(x_1, \ldots, x_n)$ is the n-th mixed Cayley–Hamilton polynomial.

Theorem 6.1.5. (i) (Procesi [Pr3]) *The following isomorphisms hold:*

$$K^+\langle X\rangle/(x^n)^T \cong K^+\langle T(X), X\rangle/(\Psi_n(x_1,\ldots,x_n), \operatorname{tr}(y_1\cdots y_m), \ m \geq 1)^T$$
$$\cong T_n^+/(\operatorname{tr}(y_{i_1}\cdots y_{i_m}) \mid y_{i_p} \in Y, m \geq 1),$$

where $(x^n)^T \subset K^+\langle X\rangle$, $(\Psi_n(x_1,\ldots,x_n), \operatorname{tr}(y_1\cdots y_m), \ m \geq 1)^T$ *is the T-ideal of* $K^+\langle T(X), X\rangle$ *generated by the Cayley–Hamilton polynomial and all formal traces and* $(\operatorname{tr}(y_{i_1}\cdots y_{i_m}) \mid y_{i_p} \in Y, m \geq 1)$ *is the ideal of the mixed trace generic matrix algebra generated by all traces.*

(ii) (Razmyslov [Ra3], Procesi [Pr3]) *All ordinary polynomial identities of* $M_n(K)$ *follow from the polynomial identity* $x^n = 0$.

The class of nilpotency $d(n)$ in the Nagata–Higman theorem is related in the following nice way to the invariant theory of matrices.

Theorem 6.1.6. *Let* $d(n)$ *be the class of nilpotency in the Nagata–Higman theorem, let* $e(n)$ *be the least positive integer such that* T_n *is generated as a* C_n-*module by products* $y_{j_1}\cdots y_{j_q}$ *of length* $\leq e(n)$, *and let* $f(n)$ *be the least integer such that* C_n *is generated as a* K-*algebra by traces* $\operatorname{tr}(y_{i_1}\cdots y_{i_m})$ *of degree* $\leq f(n)$. *Then*

$$d(n) = e(n) + 1 = f(n).$$

Proof. First we shall establish the "easy" inequalities $e(n) \leq d(n) - 1$ and $f(n) \leq d(n)$. By Corollary 6.1.3, for any $q \geq d = d(n)$, the product $x_{j_1}\cdots x_{j_q}$ has the form

$$x_{j_1}\cdots x_{j_q} = \sum \alpha_u e_n(u_1,\ldots,u_n) = \sum \alpha_u \sum_{\sigma \in S_n} u_{\sigma(1)}\cdots u_{\sigma(n)},$$

where $\alpha_u \in K$ and $u_1 = u_1(X),\ldots, u_n = u_n(X)$ are monomials of positive degree in the set of free variables X. The total degree of the product $u_1\cdots u_n$ is equal to q. Hence

$$y_{j_1}\cdots y_{j_q} = \sum \alpha_u e_n(u_1(Y),\ldots,u_n(Y)).$$

Applying the linearization of the Cayley–Hamilton theorem, we have the equality $\Psi_n(u_1(Y),\ldots,u_n(Y)) = 0$ and we may replace each $e_n(u_1(Y),\ldots,u_n(Y))$ with linear combination of shorter products $u_{k_1}(Y)\cdots u_{k_p}(Y)$, $0 \leq p \leq n - 1$, with coefficients which are products of traces $\operatorname{tr}(u_{i_1}(Y)\cdots u_{i_m}(Y))$. By induction we obtain that $y_{j_1}\cdots y_{j_q}$ is a linear combination of $y_{k_1}\cdots y_{k_p}$, with $0 \leq p \leq d(n) - 1$, and with coefficients from C_n.

By the inequality $e(n) \leq d(n) - 1$ we obtain that

$$y_{i_1}\cdots y_{i_d} = \sum_{q \leq d-1}\sum_j c_j y_{j_1}\cdots y_{j_q}, \quad c_j \in C_n.$$

Clearly, the coefficients c_j are polynomials in $\operatorname{tr}(y_{k_1}\cdots y_{k_p})$ with $p \leq d$. Multiplying by $y_{i_{d+1}}$ and taking the trace, we obtain

$$\operatorname{tr}(y_{i_1}\cdots y_{i_d}y_{i_{d+1}}) = \sum_{d \leq d-1}\sum_j c_j \operatorname{tr}(y_{j_1}\cdots y_{j_q}y_{i_{d+1}}),$$

and $\operatorname{tr}(y_{i_1} \cdots y_{i_d} y_{i_{d+1}})$ is expressed in terms of $\operatorname{tr}(y_{j_1} \cdots y_{j_q})$ with $q \leq d$. The general case $m > d(n)$ follows easily by induction. By the inequality $e(n) \leq d(n)-1$ again,

$$y_{i_1} \cdots y_{i_{m-1}} = \sum_{q \leq d-1} \sum_j c_j y_{j_1} \cdots y_{j_q},$$

where the coefficients c_j are polynomials in the traces $\operatorname{tr}(y_{j_1} \cdots y_{j_q})$ with $q \leq d$. Hence

$$\operatorname{tr}(y_{i_1} \cdots y_{i_m} y_{i_{d+1}}) = \sum_{q \leq d-1} \sum_j c_j \operatorname{tr}(y_{j_1} \cdots y_{j_q} y_{i_m}),$$

and this completes the proof of the inequality $f(n) \leq d(n)$.

For the proofs of the opposite inequalities $e(n) \geq d(n) - 1$ and $f(n) \geq d(n)$ we have to apply Theorem 6.1.5. □

6.2 Bounds for the Class of Nilpotency

In the previous section we gave some arguments that it is important to know the exact value of the class of nilpotency $d = d(n)$ in the Nagata–Higman theorem. The upper bound given in the proof of Higman [Hg], see Remark 6.1.4, is $d(n) \leq 2^n - 1$. The best known upper bound is due to Razmyslov [Ra3]. Applying trace polynomial identities of matrices, as a consequence of the theorem that the polynomial identities of $M_n(K)$ follow from $x^n = 0$, he obtained the bound $d(n) \leq n^2$. The proof of the theorem of Razmyslov may be found also in his book [Ra5] or in the book by Formanek [F6]. For a lower bound, Kuzmin [Ku] showed that $d(n) \geq \frac{1}{2}n(n + 1)$. Hence

$$\frac{n(n + 1)}{2} \leq d(n) \leq n^2.$$

Problem 6.2.1. *Find the exact value $d(n)$ of the class of nilpotency of nil algebras of index n (over a field of characteristic 0).*

Conjecture 6.2.2. (Kuzmin [Ku]) *The exact value $d(n)$ of the class of nilpotency of nil algebras of index n over a field of characteristic 0 is*

$$d(n) = \frac{n(n + 1)}{2}.$$

The only values of $d(n)$ are known for $n \leq 4$: Dubnov [Du] obtained in 1935

$$d(1) = 1, d(2) = 3, d(3) = 6.$$

In 1993, Vaughan–Lee [VL] proved that

$$d(4) = 10.$$

In this way the conjecture of Kuzmin is confirmed for $n \leq 4$.

Now we present the proof of the theorem of Kuzmin [Ku] for the lower bound of the class of nilpotency. The idea is to show that the monomial in two variables

$$u = yxyx^2y \cdots x^{n-2}yx^{n-1} \qquad (6.2)$$

does not belong to the T-ideal generated by x^n.

We shall work in the free nonunitary algebra $K^+\langle x, y \rangle$. Let us denote by J the T-ideal of $K^+\langle x, y \rangle$ generated by x^n (i.e., J is the set of all consequences of $x^n = 0$ which can be written in two variables only).

We introduce some notation:

$A^{(k)}$ is the vector space spanned by all monomials

$$u_a = x^{a_1}yx^{a_2}y \cdots x^{a_{k-1}}yx^{a_k} \qquad (6.3)$$

of degree $k - 1$ with respect to y, $1 \leq k \leq n$;

$B^{(k)}$ is the subspace of $A^{(k)}$ spanned by all polynomials of the following three kinds:

(1) u_a from (6.3) such that $a_i = a_j$ for some different indices i, j;
(2) u_a from (6.3) such that $a_i \geq n$ for some index i;
(3) the sums

$$x^{a_1}y \cdots x^{a_i} \cdots x^{a_j} \cdots yx^{a_k} + x^{a_1}y \cdots x^{a_j} \cdots x^{a_i} \cdots yx^{a_k}$$

for some different indices i, j.

We denote $A^{(n)}$ by A and $B^{(n)}$ by B.

Finally, C is the vector subspace of A spanned by all

$$x^{\sigma(0)}yx^{\sigma(1)}y \cdots x^{\sigma(n-2)}yx^{\sigma(n-1)},$$

where $\sigma \in S_n$ and S_n acts on the set $\{0, 1, 2, \ldots, n-1\}$.

We shall prove that $C \cap J \subset B$. Since the monomial (6.2) is of degree

$$(0 + 1 + 2 + \cdots + (n-1)) + (n-1) = \frac{n(n+1)}{2} - 1$$

and does not belong to B this will give that $d(n) \geq n(n+1)/2$. The proof will be carried out following the main steps of the original exposition of Kuzmin [Ku] but, for simplicity of notation, we prefer to translate some parts of the proof in terms of commutative algebra, as in the Formanek construction of central polynomials [F1]. We consider the bijection φ_k between the polynomials in the commutative variables $t_1, \ldots, t_k$ and the elements of $A^{(k)}$ defined by

$$\varphi_k : t_1^{a_1}t_2^{a_2} \cdots t_{k-1}^{a_{k-1}}t_k^{a_k} \rightarrow x^{a_1}yx^{a_2}y \cdots x^{a_{k-1}}yx^{a_k}.$$

We denote $A_0^{(k)} = K[t_1, \ldots, t_k]$ and $J_0^{(k)} = \varphi_k^{-1}(J \cap A^{(k)})$, $B_0^{(k)} = \varphi_k^{-1}(B^{(k)})$, $A_0 = A_0^{(n)}$, $\varphi = \varphi_n$, $J_0 = J_0^{(n)}$, $B_0 = B_0^{(n)}$, $C_0 = \varphi^{-1}(C)$. It is clear that for any polynomial $f(t_1, \ldots, t_{k-1}) \in B_0^{(k-1)}$ and any c_1, the polynomial $t_1^{c_1} f(t_2, \ldots, t_k)$ belongs to $B_0^{(k)}$. Finally, for any integers $k \leq n$ and $p \geq 1$, let

$$h_{k,p}(t_1, \ldots, t_k) = \sum t_1^{c_1} \cdots t_k^{c_k}, \tag{6.4}$$

where the sum is over all k-tuples $(c_1, \ldots, c_k)$ such that $c_1 + \cdots + c_k = p$.

Lemma 6.2.3. *Let*

$$u = u(t_1, \ldots, t_k) = t_1^{a_1} t_2^{a_2} \cdots t_k^{a_k}$$

be a monomial such that $0 \leq a_1 < a_2 < \cdots < a_k \leq n - 1$ and let $a_1 + k + p > n$. Then

$$h_{k,p}(t_1, \ldots, t_k) u(t_1, \ldots, t_k) \in B_0^{(k)}.$$

Proof. We use induction on k. If $k = 1$, then $u = t_1^{a_1}$, $0 \leq a_1 \leq n - 1$, $h_{1,p} = t_1^p$, $h_{1,p} u = t_1^{a_1 + p}$ and $a_1 + k + p = a_1 + 1 + p > n$. Hence $a_1 + p \geq n$ and

$$h_{1,p}(t_1) u(t_1) = t_1^{n+q}$$

for some $q \geq 0$. In this way $h_{1,p}(t_1) u(t_1)$ belongs to $B_0^{(1)}$.

Now, we assume that $k > 1$. Then

$$h_{k,p} u = \sum_{\sum c_i = p} t_1^{a_1 + c_1} t_2^{a_2 + c_2} \cdots t_k^{a_k + c_k}. \tag{6.5}$$

We consider an arbitrary summand

$$t_1^{a_1 + d_1} t_2^{a_2 + d_2} \cdots t_k^{a_k + d_k} \tag{6.6}$$

of $h_{k,p} u$. If $0 \leq d_1 < a_2 - a_1$, then the sum of all monomials in (6.5) with $c_1 = d_1$ is

$$\sum_{\sum c_i = p - d_1} t_1^{a_1 + d_1} t_2^{a_2 + c_2} \cdots t_k^{a_k + c_k} = t_1^{a_1 + d_1} h_{k-1, p - d_1}(t_2, \ldots, t_k) t_2^{a_2} \cdots t_k^{a_k}.$$

Since $d_1 < a_2 - a_1$, we obtain that $a_1 + d_1 < a_2$, hence $a_1 + d_1 + 1 \leq a_2$,

$$n < a_1 + k + p = a_1 + k + (d_1 + c_2 + \cdots + c_k)$$
$$= (a_1 + d_1 + 1) + (k - 1) + (c_2 + \cdots + c_k) \leq a_2 + (k - 1) + p',$$

where $p' = c_2 + \cdots + c_k$. By inductive arguments, since $a_2 + (k - 1) + p' > n$, we obtain that $h_{k-1, p-d_1}(t_1, \ldots, t_{k-1}) t_1^{a_2} \cdots t_{k-1}^{a_k} \in B_0^{(k-1)}$. Hence

$$h_{k,p} u = t_1^{a_1 + d_1} h_{k-1, p - d_1}(t_2, \ldots, t_k) t_2^{a_2} \cdots t_k^{a_k} \in B_0^{(k)}.$$

Now, let the summand (6.6) satisfy $d_1 \geq a_2 - a_1$. Hence $d_1 = a_2 - a_1 + i$ and $d_2 = j$ for some $i, j \geq 0$. Consider the summand

$$t_1^{a_1 + e_1} t_2^{a_2 + e_2} t_3^{a_3 + d_3} \cdots t_k^{a_k + d_k}, \tag{6.7}$$

where $e_1 = a_2 - a_1 + j$ and $e_2 = i$. If $i = j$, then $d_1 = e_1$, $d_2 = e_2$ and the expressions (6.6) and (6.7) are both equal to

$$t_1^{a_1+(a_2-a_1+i)} t_2^{a_2+i} t_3^{a_3+d_3} \cdots t_k^{a_k+d_k} = t_1^{a_2+i} t_2^{a_2+i} t_3^{a_3+d_3} \cdots t_k^{a_k+d_k}$$

and the summand (6.6) belongs to $B_0^{(k)}$. If $i \neq j$, then the summands (6.6) and (6.7) are different and their sum is equal to

$$(t_1^{a_2+i} t_2^{a_2+j} + t_1^{a_2+j} t_2^{a_2+i}) t_3^{a_3+d_3} \cdots t_k^{a_k+d_k}$$

and hence belongs to $B_0^{(k)}$. In this way, we split the expression $h_{k,p}u$ in three parts, each of them belonging to $B_0^{(k)}$: the summands with fixed $d_1 < a_2 - a_1$, the summands with $d_1 = a_2 - a_1 + i$ and $d_2 = i$ and pairs of monomials with $d_1 = a_2 - a_1 + i$, $d_2 = j$ and $e_1 = a_2 - a_1 + j$, $e_2 = i$ with $i \neq j$. This gives that $h_{k,p}u$ also belongs to $B_0^{(k)}$. $\square$

Lemma 6.2.4. Let $u = u(t_1, t_2, \ldots, t_k) = t_1^{a_1} t_2^{a_2} \cdots t_k^{a_k}$ be any monomial and let $k + p > n$. Then

$$h_{k,p}(t_1, \ldots, t_k) u(t_1, \ldots, t_k) \in B_0^{(k)}.$$

Proof. First, let

$$u(t_1, \ldots, t_k) = t_1^{a_1} t_2^{a_2} \cdots t_k^{a_k} \in B_0^{(k)}.$$

This means that $a_i \geq n$ for some i or $a_i = a_j$ for some different i and j. If $a_i \geq n$, then the same inequality $a_i + c_i \geq n$ holds for each summand of

$$h_{k,p}u = \sum_{\sum c_i = p} t_1^{a_1+c_1} t_2^{a_2+c_2} \cdots t_k^{a_k+c_k}$$

and $h_{k,p}u \in B_0^{(k)}$. If $a_i = a_j$ for some $i < j$, then we divide the summands of $h_{k,p}u$ in two groups. The first group consists of all

$$t_1^{a_1+c_1} \cdots t_i^{a_i+c_i} \cdots t_j^{a_i+c_i} \cdots t_k^{a_k+c_k}$$

with $c_i = c_j$ and these monomials belong to $B_0^{(k)}$. We split all other summands in pairs

$$t_1^{a_1+c_1} \cdots t_i^{a_i+c_i} \cdots t_j^{a_i+c_j} \cdots t_k^{a_k+c_k} + t_1^{a_1+c_1} \cdots t_i^{a_i+c_j} \cdots t_j^{a_i+c_i} \cdots t_k^{a_k+c_k}$$

with $c_i < c_j$. Again these sums are in $B_0^{(k)}$ and hence $h_{k,p}u \in B_0^{(k)}$. If

$$u(t_1, \ldots, t_k) = t_1^{a_1} t_2^{a_2} \cdots t_k^{a_k} \notin B_0^{(k)},$$

then all a_i's are pairwise different and $< n$. Since $a_1, \ldots, a_k$ behave symmetrically, we may assume that $0 \leq a_1 < a_2 < \cdots < a_k \leq n-1$. The condition $a_1 + k + p > n$ follows automatically from $k + p > n$ and the proof is completed by Lemma 6.2.3. $\square$

Lemma 6.2.5. *Let $e_n(x_1, \ldots, x_n)$ be the complete linearization (6.1) of the nil polynomial z^n and let $w_1, \ldots, w_{k-1}$, $k \le n$, be monomials in x, y essentially depending on y. Let $\deg_y w_1 + \cdots + \deg_y w_{k-1} = n - 1$. Then the sum*

$$e_n(w_1, \ldots, w_{k-1}, \underbrace{x, \ldots, x}_{n-k+1 \text{ times}}) \tag{6.8}$$

belongs to the vector space B.

Proof. Let us denote $p = n - k + 1$. We have the equality

$$e_n(w_1, \ldots, w_{k-1}, \underbrace{x, \ldots, x}_{p \text{ times}}) = p! \sum_j \sum_{\sum c_i = p} x^{c_1} w_{j_1} x^{c_2} w_{j_2} \cdots w_{j_{k-1}} x^{c_k},$$

where the outer sum is on all permutations $j_1, j_2, \ldots, j_{k-1}$ of $1, 2, \ldots, k-1$ and the inner sum is on all k-tuples $c_1, c_2, \ldots, c_k$ with sum equal to p. We shall show that each inner sum belongs to B. For this purpose, we write each w_j in the form $w_j = x^{b'_j} y \cdots y x^{b''_j}$. Then the sum under consideration has the form

$$
\begin{aligned}
v_j &= \sum_{\sum c_i = p} x^{c_1} w_{j_1} x^{c_2} w_{j_2} \cdots w_{j_{k-1}} x^{c_k} \\
&= \sum_{\sum c_i = p} x^{a_1 + c_1} y \cdots y x^{a_2 + c_2} y \cdots y x^{a_3 + c_3} \cdots x^{a_{k-1} + c_{k-1}} y \cdots y x^{a_k + c_k},
\end{aligned}
$$

where

$$a_1 = b'_{j_1}, a_2 = b''_{j_1} + b'_{j_2}, \ldots, a_{k-1} = b''_{j_{k-2}} + b'_{j_{k-1}}, a_k = b''_{j_{k-1}}.$$

If the degree of some w_j with respect to y is ≥ 2, then maybe $w_j = x^{b'_j} y \cdots y x^{b''_j}$ contains some x'es between the y's. Hence, using the bijection $\varphi : K[t_1, \ldots, t_n] = A_0 \to A$, we obtain that

$$\varphi^{-1}(v_j) = v' \sum_{\sum c_i = p} t_1^{a_1 + c_1} t_{m_2}^{a_2 + c_2} \cdots t_{m_{k-1}}^{a_{k-1} + c_{k-1}} t_n^{a_k + c_k},$$

where $v' = t_{q_1}^{b_1} \cdots t_{q_{n-k}}^{b_{n-k}}$, $b_i \ge 0$, and the set of indices $\{q_1, \ldots, q_{n-k}\}$ is the complement of $\{m_1 = 1, m_2, \ldots, m_{k-1}, m_k = n\}$ to the whole set $\{1, 2, \ldots, n\}$. Hence

$$\varphi^{-1}(v_j) = v'(t_{q_1}, \ldots, t_{q_{n-k}}) h_{k,p}(t_{m_1}, \ldots, t_{m_k}) u(t_{m_1}, \ldots, t_{m_k}).$$

By Lemma 6.2.4, $h_{k,p}(t_1, \ldots, t_k) u(t_1, \ldots, t_k) \in B_0^{(k)}$. Since $v' = v'(t_{q_1}, \ldots, t_{q_{n-k}})$ and $h_{k,p} u = h_{k,p}(t_{m_1}, \ldots, t_{m_k}) u(t_{m_1}, \ldots, t_{m_k})$ depend on sets of variables with trivial intersection, a similar inclusion

$$\varphi^{-1}(v_j) = v' h_{k,p} u \in B_0$$

holds and this completes the proof. $\square$

Now we shall establish the main step in the proof of the theorem of Kuzmin. Recall that the linear operator δ of the algebra R is called a *derivation*, if

$$\delta(uv) = \delta(u)v + u\delta(v) \quad \text{for all} \quad u, v \in R.$$

If $\delta_0 : X \to K^+\langle X \rangle$ is any mapping, then there exists a unique derivation δ of $K^+\langle X \rangle$ which extends δ_0.

Proposition 6.2.6. *Let C be the vector space spanned by all*

$$x^{\sigma(0)} y x^{\sigma(1)} y \cdots x^{\sigma(n-2)} y x^{\sigma(n-1)},$$

where $\sigma \in S_n$ and S_n acts on the set $\{0, 1, 2, \ldots, n-1\}$, and let $J = (x^n)^T$. Then $C \cap J \subset B$.

Proof. By Corollary 6.1.3 J is spanned by all polynomials $e_n(u_1, \ldots, u_n)$ from (6.1). Since we are interested in the intersection of J with C, we may consider only these $e_n(u_1, \ldots, u_n)$ which belong to the vector space A, i.e., $u_1, \ldots, u_n$ are monomials in x and y of total degree $n-1$ in y. For every positive integer a we define a derivation δ_a of $K^+\langle x, y \rangle$ by

$$\delta_a(x) = x^a, \quad \delta_a(y) = 0.$$

Obviously, $\delta_a(A) \subseteq A$ because δ_a does not change the degree of y in the homogeneous with respect to y polynomial. Since $\delta_a(x^c) = cx^{c+a-1}$ and $a \geq 1$, direct calculations show that δ_a sends the polynomials of types (1), (2) and (3) in the definition of B to linear combinations of polynomials of the same types. Hence $\delta_a(B) \subseteq B$. Finally,

$$\delta_a(e_n(u_1, \ldots, u_n)) = \sum_{j=1}^{n} e_n(u_1, \ldots, \delta_a(u_j), \ldots, u_n)$$

and $\delta_a(J) \subseteq J$.

The elements $e_n(u_1, \ldots, u_n)$ which we consider are of the form

$$e_n(w_1, \ldots, w_{k-1}, x^{c_1}, \ldots, x^{c_p}),$$

where $w_1, \ldots, w_{k-1}$ are monomials which essentially depend on y, $p = n-k+1$ and $k \leq n$. The element $\delta_a(w_j)$ is a linear combination of monomials all essentially depending on y and $\delta_a(x^c) = cx^{c+a-1}$. Since $c \geq 1$ and $a \geq 1$ we obtain that $\delta_a(e_n(w_1, \ldots, w_{k-1}, x^{c_1}, \ldots, x^{c_p}))$ is a linear combination of elements of the same form. By Lemma 6.2.5

$$e_n(w_1, \ldots, w_{k-1}, \underbrace{x, \ldots, x}_{p \text{ times}}) \in B.$$

Hence

$$\delta_a(e_n(w_1, \ldots, w_{k-1}, x, \ldots, x)) = \sum_{j=1}^{k-1} e_n(w_1, \ldots, \delta_a(w_j), \ldots, w_{k-1}, x, \ldots, x)$$
$$+ p e_n(w_1, \ldots, w_{k-1}, x^a, x, \ldots, x) \in B.$$

Every $e_n(w_1, \ldots, \delta_a(w_j), \ldots, w_{k-1}, x, \ldots, x)$ belongs to B and we obtain that

$$e_n(w_1, \ldots, w_{k-1}, x^a, x, \ldots, x) \in B$$

for all positive integers a. Now

$$\delta_b(e_n(w_1, \ldots, w_{k-1}, x^a, x, \ldots, x))$$
$$= \sum_{j=1}^{k-1} e_n(w_1, \ldots, \delta_b(w_j), \ldots, w_{k-1}, x^a, x, \ldots, x)$$
$$+ e_n(w_1, \ldots, w_{k-1}, x^{a+b-1}, x, \ldots, x)$$
$$+ (p-1)e_n(w_1, \ldots, w_{k-1}, x^a, x^b, x, \ldots, x) \in B.$$

Again, the first sum and $e_n(w_1, \ldots, w_{k-1}, x^{a+b-1}, x, \ldots, x)$ belong to B. Hence

$$e_n(w_1, \ldots, w_{k-1}, x^a, x^b, x, \ldots, x) \in B.$$

Continuing in this way, we obtain that

$$e_n(w_1, \ldots, w_{k-1}, x^{a_1}, x^{a_2}, \ldots, x^{a_p}) \in B$$

for all positive $a_1, a_2, \ldots, a_p$. Hence $A \cap J \subseteq B$. In particular, $C \cap J \subset B$. $\square$

Theorem 6.2.7. (Kuzmin, [Ku]) *The class of nilpotency in the Nagata–Higman theorem satisfies the inequality*

$$d(n) \geq \frac{n(n+1)}{2}.$$

Proof. The monomial $u = yxyx^2y \cdots x^{n-2}yx^{n-1}$ in (6.2) is of total degree

$$d = \frac{n(n+1)}{2} - 1.$$

It is sufficient to show that it does not belong to the T-ideal of $K^+\langle x, y \rangle$ generated by z^n. Clearly, $u \in C \subset A$ and $u \notin B$. Then the proof follows immediately from Proposition 6.2.6. $\square$

Remark 6.2.8. Recently, Shestakov and Zhukavets [SZ] proved that the class of nilpotency of the 2-generated algebras satisfying the identity $x^5 = 0$ is equal to 15, which agrees with the conjecture of Kuzmin for $n = 5$. They obtained the same result also in the more general setup of 2-generated superalgebras. Their proof is based on computer calculations with the GAP package. The result of Shestakov and Zhukavets suggests the following problem.

Problem 6.2.9. *Prove the conjecture of Kuzmin for 2-generated algebras of bounded nil index.*

Chapter 7

The Shirshov Theorem for Finitely Generated PI-algebras

7.1 The Shirshov Theorem and the Kurosh Problem

Algebras with polynomial identities generalize commutative and finite dimensional algebras. This generalization is not only formal. PI-algebras enjoy many other properties of commutative and finite dimensional algebras. In this section we shall present the Shirshov theorem which, roughly speaking, states that finitely generated algebras behave as finitely generated modules of polynomial algebras. Then we shall apply it to the positive solution of the Kurosh problem for PI-algebras.

We assume that the base field K is arbitrary (and of any characteristic). We fix an integer $d > 1$ and consider the set $W_d = \langle x_1, \ldots, x_d \rangle$ of all monomials (words) in the free algebra $K\langle x_1, \ldots, x_d \rangle$. The set W_d has a natural multiplication (as in the free algebra) and is called the *free semigroup* of rank d.

Definition 7.1.1. (i) We introduce a partial lexicographic ordering on W_d assuming that $x_1 < x_2 < \cdots < x_d$, and then extending it on W_d in the following way:

$$x_{i_1} \cdots x_{i_p} > x_{j_1} \cdots x_{j_q}$$

if and only if $i_1 = j_1, \ldots, i_k = j_k, i_{k+1} > j_{k+1}$ for some $k \geq 0$. (We do not compare two words u and v if one of them is a beginning of the other, i.e., $u = vw$ (or $v = uw$) for some $w \in W_d$, $w \neq 1$.)

(ii) The word $w \in W_d$ is called *n-decomposable* if it can be written in the form

$$w = w_0 w_1 \cdots w_n w_{n+1},$$

(some of the words w_0 and w_{n+1} may be empty) and

$$w_0 w_1 \cdots w_n w_{n+1} > w_0 w_{\sigma(1)} \cdots w_{\sigma(n)} w_{n+1}$$

for every nontrivial permutation $\sigma \in S_n$.

For example,

$$w = x_2 x_1 x_3 x_2 x_1 x_2 x_3 x_1 x_2 x_4 x_1 = (x_2 x_1)(x_3 x_2 x_1 x_2)(x_3 x_1 x_2)(x_4 x_1)$$

is a 2-decomposition, with $w_0 = x_2 x_1$ and $w_3 = x_4 x_1$, because

$$(x_2 x_1)(x_3 x_2 x_1 x_2)(x_3 x_1 x_2)(x_4 x_1) > (x_2 x_1)(x_3 x_1 x_2)(x_3 x_2 x_1 x_2)(x_4 x_1)$$

and
$$w = w_0 w_1 w_2 w_3 > w_0 w_2 w_1 w_3 = w_0 w_{\sigma(1)} w_{\sigma(2)} w_3$$

for the only nontrivial permutation $\sigma = (2\ 1) \in S_2$. This word has also a 4-decomposition

$$w = (x_2 x_1)(x_3 x_2 x_1 x_2)(x_3 x_1)(x_2 x_4)(x_1),\ w_0 = x_2 x_1,\ w_5 = 1.$$

The following combinatorial lemma of Shirshov [Sh] is the main step in the proof of his theorem on finitely generated PI-algebras.

Lemma 7.1.2. *Let $d > 1$ and $n > 1$ be integers. Then there exists a positive integer h such that every word $w \in W_d$ which is not n-decomposable has the form*

$$w = u_1^{a_1} \cdots u_m^{a_m} \tag{7.1}$$

where $u_1, \ldots, u_m$ are words of length $\leq n - 1$ and $m \leq h$.

There are several expositions of this lemma and its proof in the literature. For example, a proof close to the original proof of Shirshov can be found in the book by Zhevlakov, Slinko, Shestakov and Shirshov [ZSSS]. A version of A. Ya. Belov is given in the survey article by Amitsur and Small [AS] and in the book of the author [Dr10], see also the comments on the essential height below.

The words of the form (7.1) are similar to monomials in h commuting variables. The following definition is a slight modification of the original definition suggested by Shirshov.

Definition 7.1.3. Let R be an algebra generated by $r_1, \ldots, r_d$. Let H be a finite set of words of $r_1, \ldots, r_d$. One says that R is *of height h with respect to the set H* if h is the minimal integer with the property that, as a vector space, R is spanned by all products

$$u_{i_1}^{k_1} \cdots u_{i_m}^{k_m}$$

such that $u_{i_1}, \ldots, u_{i_m} \in H$ and $m \leq h$.

For example, the height of the polynomial algebra $K[x_1, \ldots, x_d]$ with respect to the set of words $H = \{x_1, \ldots, x_d\}$ is equal to d. Hence, every commutative algebra R generated by the set $H = \{r_1, \ldots, r_d\}$ has height $\leq d$ with respect to H.

Theorem 7.1.4. (Shirshov theorem [Sh]) *Let R be a PI-algebra generated by d elements $r_1, \ldots, r_d$ and satisfying a polynomial identity of degree $n > 1$. Then R is of finite height with respect to the set of all words $r_{i_1} \cdots r_{i_k}$ of length $k < n$.*

Proof. If a PI-algebra R satisfies a polynomial identity of degree n, then it satisfies also a multilinear polynomial identity of degree $\leq n$. Hence we may assume that R satisfies an identity of the form

$$x_1 \cdots x_n = \sum_{\sigma \in S_n} \alpha_\sigma x_{\sigma(1)} \cdots x_{\sigma(n)}, \tag{7.2}$$

where $\alpha_\sigma \in K$ and the summation is on all nontrivial permutations $\sigma \in S_n$. Consider a product $w = r_{i_1} \ldots r_{i_p} \in R$. By the lemma of Shirshov, there exists a positive integer h depending on d and n with the following property. If a word w is not n-decomposable, then it is of height $\leq h$ with respect to the set of all words $r_{i_1} \cdots r_{i_k}$ of length $k \leq n-1$. If the word w is n-decomposable, then we can write it as a product of $n+2$ subwords such that

$$w = w_0 w_1 \cdots w_n w_{n+1} > w_0 w_{\sigma(1)} \cdots w_{\sigma(n)} w_{n+1}$$

for any nontrivial permutation $\sigma \in S_n$. Then we apply the polynomial identity (7.2) and obtain

$$w_0(w_1 \cdots w_n) w_{n+1} = \sum_{\sigma \in S_n} \alpha_\sigma w_0 (w_{\sigma(1)} \cdots w_{\sigma(n)}) w_{n+1}.$$

Hence w is a linear combination of words which are lower in the lexicographic ordering. Continuing the process with the summands participating in this linear combination, finally we shall obtain that all elements of R are linear combinations of words in $r_1, \ldots, r_n$, which are not n-decomposable and hence of height $\leq h$. $\quad\square$

In 1941 Kurosh [Kr] asked the following problem.

Problem 7.1.5. (Kurosh problem) *Let R be a finitely generated associative algebra such that every element of R is algebraic.*
   (i) *Is the algebra R finite dimensional? If R is nil, is it nilpotent?*
   (ii) *If every element of R is algebraic of bounded degree, is R finite dimensional? If R is nil of bounded index, is it nilpotent?*

The Kurosh problem is a ring theoretic analogue of the famous Burnside problem in group theory [Bu] from 1902 whether a finitely generated periodic group is finite, and similarly for groups of fixed exponent.

The negative answer to part (i) of the Kurosh problem was given by Golod and Shafarevich [GoS]. They used some quantitative approach to construct a series of counterexamples, which were used by Golod [Go] to produce counterexamples also to the Burnside problem for periodic groups without restriction on the exponent. Concerning part (ii) of the Kurosh problem, if the characteristic of the field is 0 or sufficiently large, the Nagata–Higman theorem gives that the algebra is nilpotent even without the condition that it is finitely generated.

The nil algebras of bounded index satisfy the polynomial identity $x^n = 0$ for some n. Similarly, if all elements of the algebra are algebraic of bounded degree n, then $1, a, a^2, \ldots, a^n$ are linearly depended for any $a \in R$ and this implies that R satisfies the identity of algebraicity, as the $n \times n$ matrix algebra. One may expect that the Kurosch problem has a positive solution for PI-algebras.

Problem 7.1.6. (Kurosch problem for PI-algebras) *If R is a finitely generated PI-algebra and every element of R is algebraic (or nil), is R finite dimensional (or nilpotent)?*

For nil PI-algebras the problem was answered in the affirmative by Levitzki [Lv] and the general problem for algebraic PI-algebras was solved, also positively, by Kaplansky [Ka1]. Both proofs involve structure theory of rings and can be found in the book by Herstein [He]. Shirshov [Sh] showed that the theorems of Levitzki and Kaplansky follow from his theorem. Here we give the result of Shirshov which generalizes essentially the original results of Levitzki and Kaplansky.

Theorem 7.1.7. *Let R be a PI-algebra satisfying a polynomial identity of degree n and generated by a finite number of elements $r_1, \ldots, r_d$. Let H be the set of all products $r_{i_1} \cdots r_{i_k}$, $k < n$.*

(i) *If every element of the set H is nil, then the algebra R is nilpotent.*

(ii) *If every element of H is algebraic, then the algebra R is finite dimensional.*

Proof. By the theorem of Shirshov, R is spanned by the products

$$w = u_{i_1}^{a_1} \cdots u_{i_k}^{a_k},$$

where k is bounded by the height h and u_{i_j} are words of length $< n$ in the generators $r_1, \ldots, r_d$. Since all possible words u_{i_j} are a finite number, there is an upper bound m for the class of their nilpotency or for the degree of their algebraicity. Hence, if all u_{i_j} are nil and the sum $a_1 + \cdots + a_k$ is sufficiently large (e.g., $> h(m-1)$), then some u_{i_j} appears of degree higher than $m - 1$ and the word is equal to 0. If the elements u_{i_j} are algebraic of degree $\leq m$, then the higher degrees of u_{i_j} can be expressed as linear combinations of $1, u_{i_j}, \ldots, u_{i_j}^{m-1}$. Hence R is spanned by all words w with $a_i < m$ and the number of these words is finite. $\square$

The original proof of the theorem of Shirshov gives an upper bound of the height in terms of the number of the generators and the degree of the polynomial identity. The best known estimate

$$h \leq \frac{(n^5 - n^4)d^n}{2}$$

is obtained from the proof of Belov, see [AS], a simlified version of this proof is given also in [Dr10].

Another possibility is to find better bounds for the length of the words used in the theorem of Shirshov. It turns out that this is closely related with polynomial identities of matrices.

Definition 7.1.8. The PI-algebra R is of *PI-degree p* (or of *complexity p*), if p is the largest integer such that all multilinear polynomial identities of R follow from the multilinear identities of $M_p(K)$.

For example, the algebra $UT_k(K)$ of $k \times k$ upper triangular matrices satisfies the polynomial identity

$$[x_1, x_2] \cdots [x_{2k-1}, x_{2k}] = 0,$$

which does not hold for 2×2 matrices. Since all identities of the algebra $M_1(K) = K$ of 1×1 matrices follow from the commutator identity $[x_1, x_2] = 0$ (over an infinite field K) we obtain that the PI-degree of $UT_k(K)$ is equal to 1. (Polynomial identities which do not hold for 2×2 matrices are called *nonmatrix identities*.)

Shestakov, see [Lvv], conjectured that the bound $n - 1$ for the length of the words u_i in (7.1) can be replaced with the bound $\lceil n/2 \rceil$. Lvov [Lvv] payed attention that this is a consequence of the following statement:

If $r_1, \ldots, r_d$ are elements of the $n \times n$ matrix algebra $M_n(K)$ over the field K and all products $r_{i_1} \cdots r_{i_s}$, $s \leq n$ are nil, then the algebra generated by $r_1, \ldots, r_d$ is nilpotent. This would also imply that the length of the u_i's in (7.1) can be bounded by the PI-degree k of R.

The affirmative answer of the conjecture of Shestakov was given by Ufnarovski [U1], Belov [B1] and Chekanu [Ck]. Ufnarovski and Chekanu established the nilpotency of the subalgebras of the matrix algebra suggested by Lvov and then derived the Shirshov theorem with the PI-degree as a bound of the products using the Razmyslov–Kemer–Braun theorem for the nilpotency of the radical of finitely generated PI-algebras (Theorem 7.1.17 below). They also gave an example which shows that the bound for the length of $u_i \leq k = \mathrm{PIdeg}\, R$ is the best possible. The approach of Belov was closer to the original approach of Shirshov [Sh] but involved also combinatorics of infinite words, see also the paper by Belov [B2]. Another proof is given in the survey article by Belov, Borisenko and Latyshev [BBL]. It also uses combinatorics of infinite words, combined with the identity of algebraicity which is specific for the matrix algebras. This article contains also different relations between the bounds for the height with respect to the words of length $< n$ and those of length $\leq k$ (where n is the degree of the polynomial identity and k is the PI-degree).

Now we give another proof of the theorem of Ufnarovski, Belov and Chekanu. It seems that our proof is more transparent than the existing proofs. In particular, it does not use the Razmyslov–Kemer–Braun theorem. Our approach follows the main steps of the original proof of Shirshov. The main difference is that we additionally make use of the polynomial identity of algebraicity (as in [BBL]). We also apply two classical results of Amitsur on PI-algebras [A1] — that the only semiprimitive T-ideals over an infinite field are the T-ideals of the matrix algebras (see also Theorem 7.1.16 below) and that the Jacobson radical of a relatively free algebra is nil. (The only case which is not covered by our result is when R satisfies an identity which does not hold for the base field K. But then R is nil and hence nilpotent, so we may choose the length of the words u_i to be equal to 1.)

We fix an integer $k \geq 1$ and consider the polynomial

$$b_k(x; y_1, \ldots, y_{k+1}) = \sum_{\sigma \in S_{k+1}} (\mathrm{sign}\ \sigma) x^{\sigma(0)} y_1 x^{\sigma(1)} y_2 \cdots x^{\sigma(k)} y_{k+1}, \qquad (7.3)$$

where the symmetric group S_{k+1} acts on the set of symbols $\{0, 1, \ldots, k\}$. Since b_k is the product of the polynomial of algebraicity $a_k(x; y_1, \ldots, y_k)$ and y_{k+1},

Lemma 3.1.7 gives that $b_k(x; y_1, \ldots, y_{k+1}) = 0$ is a polynomial identity for $M_k(K)$. Let

$$g(z_1, \ldots, z_k; y_1, \ldots, y_{k+1}) \tag{7.4}$$

be the partial linearization of (7.3) of degree i in z_i, i.e., $g(z_1, \ldots, z_k; y_1, \ldots, y_{k+1})$ is the multihomogeneous component of degree i in z_i, $i = 1, \ldots, k$, of

$$b_k(z_1 + \cdots + z_k; y_1, \ldots, y_{k+1}) \in K\langle z_1, \ldots, z_k, y_1, \ldots, y_{k+1}\rangle.$$

Obviously $g = 0$ is also a polynomial identity for $M_k(K)$.

We consider the partial lexicographic ordering introduced in Definition 7.1.1. It is easy to see that if u is not a power of its proper subword then for every decomposition $u = vw$ with $1 \neq v \neq u$ the words u and wv are different and hence comparable.

Proposition 7.1.9. *Let u be a word in $x_1, \ldots, x_d$ of length $m > k$ which is not a power of a proper subword and such that for all possible decompositions*

$$u = v_1 w_1 = v_2 w_2 = \cdots = v_{m-1} w_{m-1}$$

with $1 \neq v_i \neq u$ the inequality $u > w_i v_i$ holds. Let $w_1 v_1 > w_2 v_2 > \cdots > w_{m-1} v_{m-1}$. Then for the polynomial $g(z_1, \ldots, z_k; y_1, \ldots, y_{k+1})$ from (7.4) and

$$\bar{z}_1 = w_1 v_1, \quad \bar{z}_2 = w_2 v_2, \ldots, \quad \bar{z}_k = w_k v_k,$$

$$\bar{y}_1 = v_1, \quad \bar{y}_2 = w_1 v_2, \quad \bar{y}_3 = w_2 v_3, \ldots, \quad \bar{y}_k = w_{k-1} v_k, \quad \bar{y}_{k+1} = w_k$$

the leading term of $g^c(\bar{z}_1, \ldots, \bar{z}_k; \bar{y}_1, \ldots, \bar{y}_{k+1})$ is equal to u^{pc}, where $p = k(k+3)/2$.

Proof. Clearly

$$g(z_1, \ldots, z_k; y_1, \ldots, y_{k+1})$$
$$= \sum_{\sigma \in S_{k+1}} \sum (\text{sign } \sigma) z_{i_1} \cdots z_{i_{\sigma(0)}} y_1 z_{j_1} \cdots z_{j_{\sigma(1)}} y_2 \cdots y_k z_{l_1} \cdots z_{l_{\sigma(k)}} y_{k+1},$$

where S_{k+1} acts on $\{0, 1, \ldots, k\}$ and the inner sum is on all sequences $i_1, \ldots, i_{\sigma(0)}$, $j_1, \ldots, j_{\sigma(1)}, \ldots, l_1, \ldots, l_{\sigma(k)}$ in which i appears exactly i times for every $i = 1, \ldots, k$. The term

$$\bar{g}_0 = \bar{y}_1 \bar{z}_1 \bar{y}_2 \bar{z}_2^2 \bar{y}_3 \cdots \bar{y}_k \bar{x}_k^k \bar{y}_{k+1}$$
$$= v_1(w_1 v_1)(w_1 v_2)(w_2 v_2)^2(w_2 v_3) \cdots (w_{k-1} v_k)(w_k v_k)^k w_{k+1}$$
$$= u^p,$$

where $p = k(k + 3)/2$, corresponds to the identical permutation of S_{k+1} and $j_1 = 1, \ldots, l_1 = \cdots = l_k = k$. Let $\alpha \bar{g}_1$, $\alpha \in K$, be the leading term of $\bar{g} = g(\bar{z}_1, \ldots, \bar{z}_k; \bar{y}_1, \ldots, \bar{y}_{k+1})$, where

$$\bar{g}_1 = \bar{z}_{i_1} \cdots \bar{z}_{i_{\sigma(0)}} \bar{y}_1 \bar{z}_{j_1} \cdots \bar{z}_{i_{\sigma(1)}} \bar{y}_2 \cdots \bar{y}_k \bar{z}_{l_1} \cdots \bar{z}_{l_{\sigma(k)}} \bar{y}_{k+1}.$$

Since $\bar{y}_1\bar{z}_1 = v_1(w_1v_1) = uv_1$, the beginning of $\bar{g}_0$, is bigger than $\bar{z}_{i_1} = w_{i_1}v_{i_1}$, we obtain that both $\sigma(0) = 0$ and $\bar{g}_0$ and $\bar{g}_1$ start with the same $\bar{y}_1$. Now $\sigma(1) \geq 1$ and $\bar{z}_1 > \bar{z}_i$ for $i > 1$. Therefore $j_1 = 1$. Let us assume that $\sigma(1) > 1$. Then $j_2 \geq 2$ because we have used the single $\bar{z}_1$ as $\bar{z}_{j_1}$. Clearly

$$\bar{y}_2\bar{z}_2 = (w_1v_2)(w_2v_2) = w_1uv_2 = w_1v_1w_1v_2 = \bar{z}_1w_1v_2,$$

which is bigger than $\bar{z}_{j_2}$ for all $j_2 \geq 2$. Since $\alpha\bar{g}_1$ is the leading term of $\bar{g}$, this is a contradiction and we obtain that $\sigma(1) = 1$. Hence both $\bar{g}_0$ and $\bar{g}_1$ have the same initial subword $\bar{y}_1\bar{z}_1\bar{y}_2$.

The proof is completed by obvious induction. If $\bar{g}_1 = \bar{y}_1\bar{z}_1\bar{y}_2\bar{z}_2^2\bar{y}_3\cdots\bar{z}_{q-1}^{q-1}\bar{y}_q\bar{t}$, then

$$\bar{t} = \bar{z}_{k_1}\cdots\bar{z}_{k_{\sigma(q)}}\bar{y}_{q+1}\cdots\bar{y}_k\bar{z}_{l_1}\cdots\bar{z}_{l_{\sigma(k)}}\bar{y}_{k+1},$$

σ acts identically on $\{0, 1, \ldots, q-1\}$ and $k_1, \ldots, k_{\sigma(q)}, \ldots, l_1, \ldots, l_{\sigma(k)} \geq q$. Since $\bar{z}_q^q > \bar{z}_{k_1}\cdots\bar{z}_{k_{\sigma(q)}}$ if $k_i > q$ for some $k_1, \ldots, k_q$, we obtain that

$$\bar{t} = \bar{z}_q^q\bar{z}_{k_{q+1}}\cdots\bar{z}_{k_{\sigma(q)}}\bar{y}_{q+1}\cdots\bar{y}_{k+1}.$$

Again $\bar{z}_{k_{q+1}} \leq \bar{z}_{q+1} < \bar{y}_{q+1}\bar{z}_{q+1}$ and $\sigma(q) = q$.

Since the leading term of $\bar{g}^c = g^c(\bar{z}_1, \ldots, \bar{z}_k; \bar{y}_1, \ldots, \bar{y}_{k+1})$ is the product of the leading terms $\bar{g}_0$ of $\bar{g}$ we derive the proposition. $\qquad\square$

Theorem 7.1.10. (Ufnarovski–Belov–Chekanu [U1, B1, Ck]) *If R is a finitely generated PI-algebra of PI-degree $\leq k$, then R is of finite height with respect to all words of length $\leq k$ of the generators of R.*

Proof. Let L be an extension of the field K. Then the K-algebras $L \otimes_K R$ and R have the same multilinear polynomial identities and without loss of generality we may assume that the base field K is infinite. Let R be generated by $r_1, \ldots, r_d$ and satisfy a multilinear polynomial identity $f = f(x_1, \ldots, x_n) = 0$, where $f \in T(M_k(K))$ and $f \notin T(M_{k+1}(K))$. We rewrite the identity $f = 0$ in the form (7.2). Let W be the T-ideal of $K\langle X\rangle$ generated by $f(x_1, \ldots, x_n)$. By the theorem of Amitsur [A1], see also [Rw1], Theorem 2.4.7 and Corollary 2.4.10, pp. 134–135, the Jacobson radical of the relatively free algebra $K\langle x_1, \ldots, x_d\rangle/(W \cap K\langle x_1, \ldots, x_d\rangle)$ coincides with

$$(T(M_k(K)) \cap K\langle x_1, \ldots, x_d\rangle)/(W \cap K\langle x_1, \ldots, x_d\rangle)$$

and is nil. Hence there exists a c such that $K\langle x_1, \ldots, x_d\rangle/(W \cap K\langle x_1, \ldots, x_d\rangle)$ (and hence R) satisfies the polynomial identity $g^c(z_1, \ldots, z_k; y_1, \ldots, y_{k+1}) = 0$, where g is from (7.4).

Let $h = h(d, n)$ be the height from the Shirshov theorem and let $w = r_{i_1}\cdots r_{i_s}$ be a word in the generators $r_1, \ldots, r_d$ of R. As in the proof in Theorem 7.1.4 there are the following two possibilities:

(i) w is n-decomposable, i.e., it can be decomposed as

$$w = w_0w_1\cdots w_nw_{n+1},$$

where for all non-identical permutations $\sigma \in S_n$

$$w_0 w_{\sigma(1)} \cdots w_{\sigma(n)} w_{n+1} < w_0 w_1 \cdots w_n w_{n+1}.$$

We apply the identity $f(x_1, \ldots, x_n) = 0$ to w and replace $w = w_0(w_1 \cdots w_n)w_{n+1}$
by

$$w_0 \left(\sum \alpha_\sigma w_{\sigma(1)} \cdots w_{\sigma(n)} \right) w_{n+1}$$

which is a linear combinations of terms which are lower than w in the lexicographic order.

(ii) w can be decomposed as $w = u_1^{a_1} \cdots u_N^{a_N}$, where $N \leq h$ and the length of the u_i's is less than n. We may assume that no u_i is a power of a proper subword. Let there exist a $u_0 = u_i$ of length $> k$ and of degree $a = a_j > ck(k + 3)/2$. Let $u_0 = u'u''$, where $u = u''u'$ is the biggest word obtained from u_0 by a cyclic permutation. Then $u_0^a = u'(u''u')^{a-1}u''$ and $a - 1 \geq pc$, $p = k(k + 3)/2$. Let $\bar{z}_1, \ldots, \bar{z}_k, \bar{y}_1, \ldots, \bar{y}_{k+1}$ be as in Proposition 7.1.9. By the equality $\bar{g}^c = g^c(\bar{z}_1, \ldots, \bar{z}_k; \bar{y}_1, \ldots, \bar{y}_{k+1}) = 0$ we replace the leading term u^{pc} of $\bar{g}^c$ by a linear combination of smaller elements.

By induction on the lexicographic order we express the word w as a linear combination of $v_1^{b_1} \cdots v_{N'}^{b_{N'}}$, where $N' \leq h$ and, if $b_j > ck(k+3)/2$, then the length of v_j is $\leq k$. Now we consider all $v_j^{b_j} = (r_{l_1} \cdots r_{l_{q_j}})^{b_j}$ such that v_j is of length $> k$ as products $r_{l_1} \cdots r_{l_{q_j}} \cdots r_{l_1} \cdots r_{l_{q_j}}$ of $\leq c(n - 1)k(k + 3)/2$ elements r_l of length 1 and set the height with respect to the words of length $\leq k$ equal to $h_1 = hc(n - 1)k(k + 3)/2$. $\square$

The following two corollaries from [U1, B1, Ck] follow directly from Theorem 7.1.10.

Corollary 7.1.11. *Let R be a PI-algebra generated by a finite set $\{r_1, \ldots, r_d\}$ and of PI-degree k. If all products $u_j = r_{j_1} \cdots r_{j_s}$ of length $s \leq k$ are algebraic, respectively nil, then R is finite dimensional, respectively nilpotent.*

Corollary 7.1.12. *Let the finitely generated algebra R satisfy a polynomial identity of degree n. Then in (7.1) we can assume that the length of u_i is $\leq n/2$.*

Proof. If R satisfies a polynomial identity of degree n then it satisfies a multilinear polynomial identity of degree n as well. Since the minimal degree of the multilinear identities for $M_k(K)$ is $2k$, we obtain that $k \leq n/2$ in Theorem 7.1.10. $\square$

The following easy example which is a slight modification of the examples in [U1, Ck] shows that the bound $\leq k$ for the length of u_i is the best possible in Corollary 7.1.11 and hence in Theorem 7.1.10.

Example 7.1.13. Let $M_k(K(t))$ be the K-algebra of the $k \times k$ matrices with entries from the field of rational functions in one variable and let R be the K-subalgebra of $M_k(K(t))$ generated by

$$r_1 = te_{12}, r_2 = te_{23}, \ldots, r_{k-1} = te_{k-1,k}, r_k = te_{k1}.$$

Then $K(t)R = M_k(K(t))$ and hence $T(R) = T(M_k(K(t)))$. All products $r_{j_1} \cdots r_{j_s}$ of length $s < k$ are of the form $t^s e_{pq}$ with $p \neq q$ and hence are nilpotent. On the other hand, the element $r_1 \cdots r_k = t^k e_{11}$ is not algebraic over K and the algebra R is infinite dimensional.

In most of the proofs of the Shirshov theorem, and more specially in the proof of the lemma of Shirshov (Lemma 7.1.2), one obtains that a word w in $x_1, \ldots, x_d$ which is not n-decomposable has the form

$$w = (u_1 \cdots u_{p_0}) v_1^{k_1} (u_{p_0+1} \cdots u_{p_1}) v_2^{k_2} \cdots v_t^{k_t} (u_{p_{t-1}+1} \cdots u_{p_t}), \qquad (7.5)$$

where $v_1, \ldots, v_t$ are words of length $< n$, the elements $u_1, \ldots, u_{p_0}, u_{p_0+1}, \ldots, u_{p_t}$ are equal to the variables $x_1, \ldots, x_d$ and are such that their total length is bounded in terms of n and d. It has turned out that for the quantitative investigation of finitely generated PI-algebras (e.g., for the Gelfand–Kirillov dimension) an upper bound of t in (7.5) is more important than the height.

Definition 7.1.14. Let R be an algebra generated by $r_1, \ldots, r_d$. Let H and D be finite sets of words of $r_1, \ldots, r_d$. One says that R is *of essential height h_{ess} with respect to the pair of sets H and D* if h_{ess} is the minimal integer with the property that, as a vector space, R is spanned by all products

$$c_0 v_{i_1}^{k_1} c_1 v_2^{k_2} \cdots c_{m-1} v_{i_m}^{k_m} c_m$$

such that $v_{i_1}, \ldots, v_{i_m} \in H$, $c_0, c_1, \ldots, c_m \in D$ (the c_i's are allowed also to be empty words) and $m \leq h_{\mathrm{ess}}$.

Example 7.1.15. Let $F_d(\mathfrak{U}_2) = K\langle x_1, \ldots, x_d \rangle / T(UT_2(K))$ be the relatively free free algebra of rank d in the variety generated by the 2×2 upper triangular matrices. One can show that, as a vector space, $F_d(\mathfrak{U}_2)$ has a basis

$$x_1^{a_1} \cdots x_d^{a_d}, \quad x_1^{a_1} \cdots x_d^{a_d} [x_{i_1}, x_{i_2}, \ldots, x_{i_k}],$$

$a_1, \ldots, a_d \geq 0$, $i_1 > i_2 \leq i_3 \leq \cdots \leq i_k$. The identity $[x_1, x_2][x_3, x_4] = 0$ easily implies that $F_d(\mathfrak{U}_2)$ is spanned by the products $x_1^{a_1} \cdots x_d^{a_d}$ and

$$x_1^{b_1} \cdots x_d^{b_d} [x_{i_1}, x_{i_2}] x_1^{c_1} \cdots x_d^{c_d}$$
$$= x_1^{b_1} \cdots x_d^{b_d} x_{i_1} x_{i_2} x_1^{c_1} \cdots x_d^{c_d} - x_1^{b_1} \cdots x_d^{b_d} x_{i_2} x_{i_1} x_1^{c_1} \cdots x_d^{c_d}$$

and the height of these elements is $2d + 2$. On the other hand, the essential height with respect to the sets $H = \{x_1, \ldots, x_d\}$ and $D = \{x_i x_j \mid i, j = 1, \ldots, d\}$ is equal to $2d$.

We shall complete this section with several theorems, already considered as classical, on finitely generated PI-algebras and their polynomial identities.

Theorem 7.1.16. (Amitsur [A1]) *Over an infinite field K the T-ideals $T(M_n(K))$ of the matrix algebras are the only prime ideals of $K\langle X \rangle$.*

The proof of the theorem of Amitsur can be found in Chapter 12 of the lecture notes of Formanek in the present book or in the book by Rowen [Rw1].

Theorem 7.1.17. (Razmyslov–Kemer–Braun [Ra4, Ke1, Br]) *The Jacobson radical of every finitely generated PI-algebra is nilpotent.*

Razmyslov [Ra4] proved that the Jacobson radical $J(R)$ of a finitely generated PI-algebra R over a field K of characteristic 0 is nilpotent if and only if R satisfies some Capelli identity. Kemer [Ke1] completed the theorem on the nilpotency of the radical in the case of characteristic 0 establishing that every finitely generated PI-algebra satisfies a Capelli identity. Finally, Braun [Br] proved the nilpotency of the radical in the most general setup when R is a finitely generated PI-algebra over a commutative noetherian algebra.

Definition 7.1.18. An algebra R over a field K is called *representable* if there exists an extension $\tilde{K}$ of the base field K such that R is isomorphic as a K-algebra of a subalgebra of the K-algebra $M_n(\tilde{K}) \cong \tilde{K} \otimes_K M_n(K)$ for some n.

If K is infinite then, by Remark 1.2.9 (ii), the K-algebras $M_n(\tilde{K})$ and $M_n(K)$ have the same polynomial identities, and every representable algebra satisfies all polynomial identities of $M_n(K)$.

Theorem 7.1.19. (Kemer [Ke4]) (i) *Over an infinite field K, for any PI-algebra R and any positive integer d the relatively free algebra*

$$F_d(\mathrm{var}\ R) = K\langle x_1, \ldots, x_d \rangle / (T(R) \cap K\langle x_1, \ldots, x_d \rangle)$$

is representable.

(ii) *Over a field K of characteristic 0, the polynomial identities of a finitely generated PI-algebra R coincide with the polynomial identities of some finite dimensional algebra R_0.*

The PI-degree of an algebra R measures the "complexity" of the polynomial identities of R. For finitely generated PI-algebras with nonmatrix polynomial identities there is a finer measure given by the polynomial identities of the upper triangular matrices. (Over an infinite field K the polynomial identities of $UT_n(K)$ follow from the identity $[x_1, x_2] \cdots [x_{2n-1}, x_{2n}] = 0$.)

Theorem 7.1.20. (Latyshev [La1]) *If R is a finitely generated PI-algebra over an infinite field K satisfying a nonmatrix polynomial identity, then R satisfies some polynomial identity of the form*

$$[x_1, x_2] \cdots [x_{2k-1}, x_{2k}] = 0.$$

The theorem of Latyshev was established before the theorem of Razmyslov–Kemer–Braun. Nowadays it is a direct consequence of the theorem of Amitsur and the nilpotency of the radical of finitely generated PI-algebras.

7.2 Growth of PI-algebras

We start this section with the introduction of growth functions and Gelfand–Kirillov dimension of arbitrary finitely generated algebras. For further reading we recommend the book by Krause and Lenagan [KL] and the surveys by Ufnarovski [U2] and the author [Dr9].

Definition 7.2.1. Let R be a finitely generated algebra and let $\{r_1, \ldots, r_d\}$ be a set of generators. Let

$$V^n = \mathrm{span}\{r_{i_1} \cdots r_{i_n} \mid i_j = 1, \ldots, d\}, \quad n = 0, 1, 2, \ldots,$$

where we assume that $V^0 = K$ if R is unitary and $V^0 = 0$ if it is not unitary. The *growth function* of R is defined by

$$g_V(n) = \dim \left(V^0 + V^1 + \cdots + V^n\right), \quad n = 0, 1, 2, \ldots.$$

The growth function of the algebra depends essentially on the fixed set of generators. For example, if $R = K[x_1, \ldots, x_d]$ is the polynomial algebra with the usual set of generators $\{x_1, \ldots, x_d\}$ which spans the vector space V_d, then

$$g_{V_d}(n) = \binom{n + d}{d},$$

the number of monomials in d variables and of degree $\leq n$. If we add to the generating set also all monomials of degree 2, and

$$W_d = \mathrm{span}\{x_i, x_i x_j \mid i, j = 1, \ldots, d\},$$

then

$$g_{W_d}(n) = \binom{2n + d}{d},$$

the number of monomials of degree $\leq 2n$.

In order to obtain some invariant of the growth, we define a partial ordering $\preceq$ and equivalence $\sim$ on the set of all functions $f : \mathbb{N} \cup \{0\} \to \mathbb{R}$ which are eventually monotone increasing and positive valued. This means that there exists an $n_0 \in \mathbb{N}$ such that $f(n_0) \geq 0$ and $f(n_2) \geq f(n_1) \geq f(n_0)$ for all $n_2 \geq n_1 \geq n_0$. We assume that $f \preceq g$ for two functions f and g if and only if there exist positive integers a and p such that for all sufficiently large n the inequality $f(n) \leq ag(pn)$ holds and $f \sim g$ if and only if $f \preceq g$ and $g \preceq f$.

For example, If $f(n)$ and $g(n)$ are polynomial functions with positive coefficients of the leading terms, then $f \preceq g$ if and only if $\deg f \leq \deg g$. If $\alpha, \beta > 0$, then $n^\alpha \sim n^\beta$ if and only if $\alpha = \beta$. The functions α^n and β^n are equivalent if and only if simultaneously either $\alpha = \beta = 1$ or $\alpha, \beta > 1$ (α^n is not eventually monotone increasing for $0 < \alpha < 1$).

If $V = \mathrm{span}\{r_1, \ldots, r_d\}$ is a generating vector space of the finitely generated algebra R, then the equivalence class $\mathcal{G}(R) = G(g_V) = \{f(n) \mid f(n) \sim g_V(n)\}$ is called the *growth* of R. The growth of R does not depend on the chosen system of generators.

Definition 7.2.2. Let R be a finitely generated algebra, with a set of generators $\{r_1, \ldots, r_d\}$, and let $g_V(n)$ be the growth function of R for $V = \operatorname{span}\{r_1, \ldots, r_d\}$. The *Gelfand–Kirillov dimension of R* is defined by

$$\operatorname{GKdim} R = \limsup_{n \to \infty}(\log_n g_V(n)) = \limsup_{n \to \infty} \frac{\log g_V(n)}{\log n}.$$

The Gelfand–Kirillov dimension does not depend on the choice of the set of generators.

For example, since the growth function of $R = K[x_1, \ldots, x_d]$ with respect to the usual generators $x_1, \ldots, x_d$

$$g(n) = \binom{n+d}{d} = \frac{(n+d)(n+d-1) \cdots (n+1)}{d!}$$

is a polynomial of degree d, we obtain that $\operatorname{GKdim} K[x_1, \ldots, x_d] = d$. The free associative algebra $K\langle x_1, \ldots, x_d \rangle$, $d > 1$, has no finite Gelfand–Kirillov dimension because its growth function (with respect to the generators $x_1, \ldots, x_d$) is

$$g(n) = 1 + d + d^2 + \cdots + d^n$$

and this function grows faster than any polynomial function.

Below we collect some elementary properties of Gelfand–Kirillov dimension.

Proposition 7.2.3. *Let R be a finitely generated associative algebra.*

(i) $\operatorname{GKdim} R = 0$ *if and only if R is a finite dimensional vector space. Otherwise $\operatorname{GKdim} R \geq 1$.*

(ii) *If S is a homomorphic image of R, then $\operatorname{GKdim} R \geq \operatorname{GKdim} S$.*

(iii) *If S is a finitely generated subalgebra of R, then $\operatorname{GKdim} R \geq \operatorname{GKdim} S$.*

(iv) *If the algebra R is commutative, and if S is a finitely generated subalgebra of R such that R is a finitely generated S-module, then $\operatorname{GKdim} R = \operatorname{GKdim} S$. The Gelfand–Kirillov dimension of R is equal to its transcendence degree, i.e., to the maximal number of algebraically independent elements.*

By the above Proposition 7.2.3 (iv), the Gelfand–Kirillov dimension of an arbitrary finitely generated commutative algebra is an integer. There are examples which show that any real $\alpha \geq 2$ can serve as the Gelfand–Kirillov dimension of an algebra R, see the book by Krause and Lenagan [KL] and the survey by Ufnarovski [U2]. The examples of Borho and Kraft [BK] satisfy the polynomial identity $[x_1, x_2][x_3, x_4][x_5, x_6] = 0$. A modification of this example is given also in [Dr10]. On the other hand, the following result of Bergman ([Bg1], see also [KL]) completes the answer to the question which numbers can be Gelfand–Kirillov dimensions.

Theorem 7.2.4. (Bergman gap theorem) *There exist no algebras with Gelfand–Kirillov dimension in the interval* $(1, 2)$.

Hence the Gelfand–Kirillov dimension of a finitely generated associative algebra can have as a value 0, 1 and any $\alpha \geq 2$.

Now we shall prove the theorem of Berele [Be1] that the finitely generated PI-algebras have finite Gelfand–Kirillov dimension. The original proof of Berele is based on another (weaker) version of the Shirshov theorem 7.1.4. A proof based on structure theory of rings can be found in the book by Krause and Lenagan [KL].

Theorem 7.2.5. (Berele [Be1]) *Every finitely generated PI-algebra is of finite Gelfand–Kirillov dimension.*

Proof. Let R be generated by $V = \mathrm{span}\{r_1, \ldots, r_d\}$ and let it satisfy a polynomial identity of degree n. By Theorem 7.1.4, there exists an integer h, the height of R, such that R is spanned by the words

$$u_{i_1}^{k_1} \cdots u_{i_t}^{k_t}, \quad t \leq h, \tag{7.6}$$

where $u_{i_1}, \ldots, u_{i_t}$ are all words of length $< n$. In the proof of the Shirshov theorem we have used only that R satisfies a multilinear polynomial identity. Hence V^m is spanned on those words (7.6) with

$$k_1 |u_{i_1}| + \cdots + k_t |u_{i_t}| = m.$$

Therefore, $V^0 + V^1 + \cdots + V^m$ is a subspace of the vector space spanned by all words (7.6) satisfying the condition $k_1 + \cdots + k_h \leq m$. Let p be the number of words of length $< n$ ($p = 1 + d + \cdots + d^{n-1}$). The number of sequences of indices $(i_1, \ldots, i_h)$ is bounded by p^h. Hence the dimension $g_V(m)$ of $V^0 + V^1 + \cdots + V^m$ is bounded by the product of the number of sequences $(i_1, \ldots, i_h)$ and the number of monomials of degree $\leq m$ in h variables,

$$g_V(m) \leq p^h \binom{m + h}{h}$$

which is a polynomial of degree h. Hence

$$\mathrm{GKdim}\ R \leq h,$$

the height of R.   □

The proof of the theorem of Berele gives the following corollary.

Corollary 7.2.6. *The Gelfand–Kirillov dimension of a finitely generated PI-algebra R is bounded by the height of R with respect to any finite set of monomials of any finite set of generators of R.*

The Gelfand–Kirillov dimension measures the polynomial identities in d variables for a PI-algebra R. Since the relatively free algebra $F_d(\mathrm{var}\ R)$ of rank d in the variety generated by R is isomorphic to the factor algebra of the free algebra $K\langle x_1, \ldots, x_d \rangle$ modulo the identities in d variables, it is important to know the exact value of the Gelfand–Kirillov dimension of $F_d(\mathrm{var}\ R)$. We shall start the discussion with the Gelfand–Kirillov dimension for the variety generated by the $n \times n$ matrix algebra. The algebra $F_d(\mathrm{var}\ M_n(K))$ is isomorphic to the algebra R_{nd} generated by d generic $n \times n$ matrices. It is contained in the mixed trace algebra T_{nd}, which is a finitely generated C_{nd}-module. Hence

$$\mathrm{GKdim}\ F_d(\mathrm{var}\ M_n(K)) \leq \mathrm{GKdim}\ T_{nd} = \mathrm{GKdim}\ C_{nd}$$
$$= \text{transc. deg.}\ (C_{nd}) = (d-1)n^2 + 1.$$

It is known, see Procesi [Pr2], that above we have equality:

$$\mathrm{GKdim}\ F_d(\mathrm{var}\ M_n(K)) = (d-1)n^2 + 1.$$

If the field K is infinite, then the relatively free algebra $F_d(\mathrm{var}\ UT_n(K))$ of the variety generated by the algebra of $n \times n$ upper triangular matrices has Gelfand–Kirillov dimension equal to dn. This algebra determines the exact value of the Gelfand–Kirillov dimension of any relatively free algebra with a nonmatrix polynomial identity in virtue of the following theorem announced by Markov [Mr].

Theorem 7.2.7. *If the algebra R satisfies a nonmatrix polynomial identity and n is the largest positive integer such that the algebra $UT_n(K)$ of upper triangular matrices satisfies all polynomial identities of R, then the Gelfand–Kirillov dimension of the relatively free algebra $F_d(\mathrm{var}\ R)$ is the same as the Gelfand–Kirillov dimension of the algebra $F_d(\mathrm{var}\ UT_n(K))$ and is equal to dn.*

The building blocks of the T-ideals over a field of characteristic 0 in the theory developed by Kemer [Ke2] are the T-ideals of the polynomial identities of matrix algebras with entries from the field K and from the Grassmann algebra E as well as the subalgebras $M_{p,q}$ of $M_n(E)$. Berele [Be3] calculated the Gelfand–Kirillov dimension of the relatively free algebras $F_d(\mathrm{var}\ M_n(E))$ and $F_d(\mathrm{var}\ M_{p,q})$.

By Theorem 7.1.19 of Kemer, the finitely generated relatively free algebras over infinite fields are representable. Markov [Mr] announced that such algebras have integer Gelfand–Kirillov dimension.

Theorem 7.2.8. *Finitely generated representable algebras over infinite fields have integer Gelfand–Kirillov dimension.*

A proof of this theorem can be found in Section 12.10 in the new edition of the book by Krause and Lenagan [KL].

Corollary 7.2.9. *The Gelfand–Kirillov dimension of any finitely generated relatively free algebra over an infinite field is an integer.*

All the quantitative information on the relatively free algebra $F_d(\mathfrak{V})$ is carried out by its Hilbert series $H(F_d(\mathfrak{V}), t)$. Hence the knowledge of the behaviour of $H(F_d(\mathfrak{V}), t)$ allows to compute the Gelfand–Kirillov dimension of $F_d(\mathfrak{V})$. Using the formula of Formanek for the Hilbert series of the product of two T-ideals as a function of the Hilbert series of the factors, see Proposition 2.1.6, Berele [Be4] calculated the Gelfand–Kirillov dimension of some relatively free algebras with T-ideals which are products of the T-ideals of $M_n(K)$, $M_n(E)$ and $M_{p,q}$.

In commutative algebra the Hilbert series of finitely generated objects are rational functions. Something similar happens with relatively free algebras. The author [Dr4], using purely combinatorial arguments, showed that the Hilbert series of finitely generated relatively free algebras with nonmatrix polynomial identities have rational Hilbert series. Later this result was extended by Belov [B3] for all relatively free algebras. The proof of Belov uses essentially the theory of Kemer.

Theorem 7.2.10. (Belov [B3]) *The Hilbert series of any finitely generated relatively free algebra over an infinite field is rational.*

The rationality of the Hilbert series of a finitely generated algebra does not imply that the Gelfand–Kirillov dimension is an integer, it gives only that the growth of the algebra is either polynomial or exponential. (Obvious example:

$$H(K\langle x_1, \ldots, x_d \rangle, t) = \frac{1}{1 - dt}$$

is a rational function but GKdim $K\langle x_1, \ldots, x_d \rangle = \infty$.) Nevertheless, combined with the theorem of Berele, the theorem of Belov gives that GKdim $F_m(\mathfrak{U})$ is an integer for any nontrivial variety $\mathfrak{U}$.

Repeating the main steps of the proof of the theorem of Berele, but replacing the usual height in the theorem of Shirshov with the essential height, we can easily see that the Gelfand–Kirillov dimension of a finitely generated PI-algebra R is bounded from above by the essential height. For the case of representable algebras these numbers coincide.

Theorem 7.2.11. (Belov, Borisenko, Latyshev [BBL] Theorem 2.110, p. 3505) *Let R be a representable algebra generated by a finite set $\{r_1, \ldots, r_d\}$. Let H and D be finite sets of words of $r_1, \ldots, r_d$ and let $h_{\mathrm{ess}}(R)$ be the essential height of R with respect to the pair H and D. Then*

$$\mathrm{GKdim}\ R = h_{\mathrm{ess}}(R).$$

Since the finitely generated relatively free algebras over infinite fields are representable, this immediately gives:

Corollary 7.2.12. *The Gelfand–Kirillov dimension of any finitely generated relatively free algebra $F_d(\mathfrak{U})$ of a variety $\mathfrak{U}$ of associative PI-algebras over an infinite field coincides with the essential height of $F_d(\mathfrak{U})$ with respect to any finite sets H and D of words in the generators of $F_d(\mathfrak{U})$.*

Chapter 8

Growth of Codimensions of PI-algebras

8.1 The Theorem of Regev for the Tensor Product of PI-algebras

The theorem of Shirshov in the previous chapter can be considered as quantitative expression of the fact that the class of finitely generated PI-algebras enjoys the most important properties of commutative and of finite dimensional algebras. One of the goals of this chapter is to give one more evidence that PI-algebras have nice properties. We consider multilinear polynomial identities over any field of arbitrary characteristic. The main quantitative result of the chapter is the theorem of Regev for the exponential growth of the codimension sequence of a PI-algebra: For every PI-algebra R there exists an $a > 0$ such that $c_m(R) = \mathcal{O}(a^m)$. (This is much smaller than $\dim P_m = m!$.) As a consequence we give another important theorem of Regev that the tensor product of two PI-algebras is again PI. This shows that the class of PI-algebras is closed under natural algebraic operations.

We give the proof of Latyshev [La2] based on the Dilworth theorem in combinatorics [Di]. We include the original proof of the Dilworth theorem, which, in its complete generality can be found also in the book by Hall [H]. A version of the Dilworth theorem used in the expositions of other texts is proposed by Amitsur, see [Rw1] or [Dr10].

Recall that a set P with a relation $\prec$ is *partially ordered* if the relation is transitive, (i.e., $a \prec b$ and $b \prec c$ for $a, b, c \in P$ imply $a \prec c$) and if there are no elements $a, b \in P$ such that both the relations $a \prec b$ and $b \prec a$ hold. Two elements $a, b \in P$ are *comparable*, if either $a \prec b$ or $b \prec a$. A *chain* is a sequence of elements $a_1, \ldots, a_k$ in P such that $a_1 \prec \cdots \prec a_k$; an *antichain* is a set of pairwise incomparable elements.

Theorem 8.1.1. (Dilworth [Di]) *Let $(P, \prec)$ be a partially ordered set. Then the minimal number of chains of P into which P can be presented as a disjoint union is equal to the maximum number of elements in an antichain of P.*

Proof. We shall establish the theorem only in the case of a finite set P. For the general case see the original proof [Di] or [H]. Let P be presented as a disjoint union of d chains and let d be the minimal number with this property. If $a_1, \ldots, a_e$ are pairwise incomparable elements in P, then two elements a_p and a_q cannot belong to the same chain. Hence $d \geq e$ and we have to show the opposite inequality.

We apply induction on the number m of the elements of P. We shall write $P = C_1 + \cdots + C_d$, if P is a disjoint union of the chains $C_1, \ldots, C_d$. We call the elements of a subset I independent, if any two elements of I are incomparable, i.e., I is an antichain. Hence, the proof will be completed, if we establish: If k is the maximal number of independent elements of P, then P is a sum of k chains. Let a be an arbitrary element of P. We remove a from P and obtain a new partially ordered set P^* with $m-1$ elements. Clearly, the maximal number of independent elements in P^* is equal either to $k-1$ or to k. By inductive arguments, P^* is a sum of $k-1$ or k chains, respectively. If $P^* = C_1 + \cdots + C_{k-1}$, then we build a new chain C_k consisting of the single element a and obtain $P = C_1 + \cdots + C_{k-1} + C_k$ and this completes the proof in this case. So, we may assume that

$$P - \{a\} = P^* = C_1 + \cdots + C_k.$$

Each chain C_p is a union of the three subsets B_p, S_p and I_p consisting respectively of all elements $b_p, s_p, i_p \in C_p$ such that $a \prec b_p$ (b_p is bigger than a), $s_p \prec a$ (s_p is smaller than a) and i_p and a are incomparable, respectively. Clearly, $s_p \prec i_p \prec b_p$ for $s_p \in S_p$, $i_p \in I_p$ and $b_p \in B_p$. We denote

$$B = B_1 + \cdots + B_k, \quad S = S_1 + \cdots + S_k, \quad I = I_1 + \cdots + I_k.$$

Pay attention that for each p and q, the set $S_p \cup \{a\} \cup B_q$ is a chain. Hence, if some of the sets I_p is empty, then $C_p' = S_p + \{a\} + B_p$ is a chain, $P = C_1 + \cdots + C_p' + \cdots + C_k$ and P is a sum of k chains. In this way, we may assume that all sets I_p are nonempty.

The next step in the proof is to show that for some p the maximal number of independent elements in $I + B - B_p$ is less than k. Let us assume that for all $p = 1, \ldots, k$, the set $I + B - B_p$ has a subset A_p of k independent elements. Since each C_q is a chain, A_p cannot contain two elements from C_q. Hence A_p contains exactly one element from C_q. For $p = q$, $I + B - B_p$ does not contain elements from B_p and this gives that A_p contains an element from I_p. Let $A = A_1 + \cdots + A_k$ (the sets A_p may have nonempty intersections). Let a_p be the minimal element of A contained in A_p. Since A_p already has an element from I_p, and all elements from I_p are smaller than the elements from B_p, we derive that $a_p \in I_p$. The $k + 1$ elements $a, a_1, \ldots, a_k$ cannot be independent. Since a is incomparable with the elements $a_1, \ldots, a_k$ (because $a_p \in I_p$), we obtain that two different a_p and a_q are comparable. Let for example $a_q \prec a_p$. Since A_p consists of k independent elements, a_q belongs to some A_q with $q \neq p$. The set A_q contains one element from each C_r, hence it contains an element d_p from C_p. Now, $d_p \in A \cap C_p$, the set $A \cap C_p$ is a subchain of C_p and a_p is the minimal element in this subchain. Hence $a_p \preceq d_p$. This, together with $a_q \prec a_p$ gives $a_q \prec d_p$. Both the elements a_q and d_p belong to the set A_q. This is impossible because A_q consists of independent elements. This contradiction shows that there exists a p such that the set $I + B - B_p$ contains less than k independent elements only. Similarly, there is a q such that the set $I + S - S_q$ cannot contain k independent elements.

We fix a subset T of independent elements of the set

$$(I + B - B_p) + (I + S - S_q) = P^* - B_p - S_q.$$

If $t_1, t_2 \in T$ are such that $t_1 \in B - B_p$ and $t_2 \in S - S_q$, then $t_2 \prec a \prec t_1$ and t_1 and t_2 are comparable which is impossible. Hence either $T \subset I + B - B_p$ or $T \subset I + S - S_q$. In both cases T has less than k elements. By inductive arguments, $P^* - B_p - S_q$ is a sum of less than k chains:

$$P^* - B_p - S_q = C_1^* + \cdots + C_{k-1}^*.$$

The set $C_k^* = S_q \cup \{a\} \cup B_p$ is a chain and

$$P = C_1^* + \cdots + C_{k-1}^* + C_k^*,$$

which completes the proof.   □

The idea for the following definition can be traced back to the Shirshov approach to combinatorics of words and its applications to PI-algebras. (Compare with the definition of n-decomposable words.)

Definition 8.1.2. For a permutation π in S_m we denote by $d(\pi)$ the largest number d for which there exist integers $1 \leq i_1 < i_2 < \cdots < i_d \leq m$ such that $\pi(i_1) > \pi(i_2) > \cdots > \pi(i_d)$. For a fixed integer $n > 1$, the permutation π in S_m is called *n-good* if $d(\pi) < n$.

In other words, for a fixed $\pi \in S_m$, we introduce a partial ordering on the set $P = \{1, 2, \ldots, m\}$ in the following way: $p \prec q$ if and only if $p < q$ and $\pi(p) < \pi(q)$. Then $d(\pi)$ is the maximum length of the antichains in $\{1, 2, \ldots, m\}$ and π is n-good if there are no antichains of length n.

For example, let $m = 6$ and

$$\pi = \begin{pmatrix} 1 & 2 & 3 & 4 & 5 & 6 \\ 5 & 3 & 2 & 4 & 1 & 6 \end{pmatrix}.$$

Then $d(\pi) = 4$, since $1 < 2 < 3 < 5$ and $\pi(1) > \pi(2) > \pi(3) > \pi(5)$ (and there are no antichains of length 5). Hence π is 5-good (and also 6-good).

Lemma 8.1.3. *The number of n-good permutations in S_m is bounded by $(n-1)^{2m}$.*

Proof. Let $\pi \in S_m$ be n-good. Consider the partial ordering in $\{1, 2, \ldots, m\}$ induced by π. Then the maximal length of the antichains in π is $\leq n - 1$ and by the Dilworth theorem 8.1.1, the set $\{1, 2, \ldots, m\}$ can be partitioned in $\leq n - 1$ chains, say $C_1, \ldots, C_k$, $k \leq n - 1$. Every chain corresponds to a sequence of pairs $(i_1, \pi(i_1)), \ldots, (i_p, \pi(i_p))$ with $i_1 < \cdots < i_p$ and $\pi(i_1) < \cdots < \pi(i_p)$. We define functions $t : \{1, 2, \ldots, m\} \to \{1, \ldots, n-1\}$ and $u : \{1, 2, \ldots, m\} \to \{1, \ldots, n-1\}$ by the rule $t(i) = j$ and $u(\pi(i)) = j$ if i is in the j-th chain C_j. The different n-good permutations correspond to different pairs of functions (t, u) and the total number of pairs of functions is $(n-1)^{2m}$.   □

Theorem 8.1.4. (Latyshev [La2]) *If R is a PI-algebra satisfying a polynomial iden-tity of degree n, then the vector space of multilinear polynomials of degree m in $K\langle X\rangle$ is spanned modulo the T-ideal $T(R)$ by the monomials $x_{\pi(1)}\cdots x_{\pi(m)}$, where the permutations $\pi \in S_m$ are n-good (i.e., $d(\pi) < n$).*

Proof. Since $T(R)$ contains a polynomial identity of degree n, it contains also a multilinear identity of degree n. Without loss of generality we may assume that R satisfies an identity of the form

$$x_n x_{n-1}\cdots x_1 = \sum_{\sigma \in S_n} \alpha_\sigma x_{\sigma(1)} x_{\sigma(2)}\cdots x_{\sigma(n)}, \quad \alpha_\sigma \in K, \tag{8.1}$$

and the summation is over all permutations σ different from $\delta \in S_n$, where

$$\delta = \begin{pmatrix} 1 & 2 & \ldots & n-1 & n \\ n & n-1 & \ldots & 2 & 1 \end{pmatrix}.$$

We shall work in $P_m(R) = P_m/(P_m \cap T(R))$, i.e., in the vector space of multilinear polynomials of degree m modulo the polynomial identities of R. Let G_n be the set of all monomials $x_{\pi(1)}\cdots x_{\pi(m)} \in P_m(R)$ such that the permutation $\pi \in S_m$ is n-good. We shall show that $P_m(R) = \operatorname{span}(G_n)$. We order the monomials in $P_m/(P_m \cap T(R))$ lexicographically assuming that

$$x_{\sigma(1)}\cdots x_{\sigma(m)} < x_{\tau(1)}\cdots x_{\tau(m)}$$

if and only if for some k

$$\sigma(1) = \tau(1), \ldots, \sigma(k) = \tau(k), \sigma(k+1) < \tau(k+1).$$

Let $h = x_{\pi(1)}\cdots x_{\pi(m)}$ be the minimal monomial which is not in $\operatorname{span}(G_n)$. Hence $d(\pi) \geq n$ and there exist $i_1 < \cdots < i_n$ with $\pi(i_1) > \cdots > \pi(i_n)$. We write h in the form

$$h = h_0 x_{\pi(i_1)} h_1 x_{\pi(i_2)} h_2 \cdots h_{n-1} x_{\pi(i_n)} h_n,$$

apply the polynomial identity (8.1) for

$$\bar{x}_n = x_{\pi(i_1)} h_1, \bar{x}_{n-1} = x_{\pi(i_2)} h_2, \ldots, \bar{x}_2 = x_{\pi(i_{n-1})} h_{n-1}, \bar{x}_1 = x_{\pi(i_n)} h_n$$

and obtain

$$h = h_0 \cdot (\bar{x}_n \bar{x}_{n-1}\cdots \bar{x}_1) = \sum_{\sigma \in S_n} \alpha_\sigma h_0 \bar{x}_{\sigma(1)} \bar{x}_{\sigma(2)}\cdots \bar{x}_{\sigma(n)}$$

$$= \sum_{\sigma \in S_n} \alpha_\sigma h_0 x_{\pi(i_{\sigma(n)})} h_{\sigma(n)} x_{\pi(i_{\sigma(n-1)})} h_{\sigma(n-1)}\cdots x_{\pi(i_{\sigma(1)})} h_{\sigma(1)}.$$

Since $\pi(i_1) > \ldots > \pi(i_n)$ and the summation is on $\sigma \neq \delta$, we obtain that all monomials in the latter sum are below than h in the lexicographic ordering and, by inductive arguments, belong to the vector subspace $\operatorname{span}(G_n)$ spanned by the set G_n corresponding to the n-good permutations. Hence h also belongs to $\operatorname{span}(G_n)$. This completes the proof of the theorem. □

Now we are ready to prove the theorem of Regev for the exponential growth of the codimensions of a PI-algebra R with the estimate obtained by Latyshev.

Theorem 8.1.5. (Regev [Re1]) *Let R be a PI-algebra with a polynomial identity of degree n. Then the sequence of the codimensions of the polynomial identities of R satisfies*

$$c_m(R) \leq (n-1)^{2m}, \quad m = 0, 1, 2, \ldots.$$

Proof. By Theorem 8.1.4, the vector space $P_m/(P_m \cap T(R))$ is spanned by the n-good words $x_{\pi(1)} \cdots x_{\pi(m)}$ (i.e., such that the permutations $\pi \in S_m$ are n-good). Hence $c_m(R)$ is bounded from above by the number of n-good permutations. Now the proof follows immediately from Lemma 8.1.3.   □

The following purely qualitative result of Regev answers in the affirmative the problem of Kaplansky whether the tensor product of two PI-algebras is again PI. It is remarkable that the proof is purely quantitative.

Theorem 8.1.6. (Regev [Re1]) *The tensor product $R = R_1 \otimes R_2$ of two PI-algebras R_1 and R_2 over any field K is again a PI-algebra.*

Proof. Let R_1 and R_2 satisfy, respectively, polynomial identities of degree n_1 and n_2. Therefore, by Theorem 8.1.5,

$$c_m(R_1) \leq (n_1 - 1)^{2m}, \quad c_m(R_2) \leq (n_2 - 1)^{2m}.$$

We choose m such that $c_m(R_1)c_m(R_2) < m!$. This is always possible because a^m grows slower than $m!$ and hence is smaller than $m!$ for large m. Let

$$\{g_i(x_1, \ldots, x_m) \mid i = 1, 2, \ldots, c' = c_m(R_1)\},$$
$$\{h_j(x_1, \ldots, x_m) \mid j = 1, 2, \ldots, c'' = c_m(R_2)\}$$

be bases of $P_m(R_1) = P_m/(P_m \cap T(R_1))$ and $P_m(R_2) = P_m/(P_m \cap T(R_2))$, respectively. For every permutation $\pi \in S_m$, we consider $x_{\pi(1)} \cdots x_{\pi(m)}$ as an element of $P_m(R_1)$ and write it as a linear combination

$$x_{\pi(1)} \cdots x_{\pi(m)} = \sum_{i=1}^{c'} \beta_{\pi,i} g_i(x_1, \ldots, x_m), \quad \beta_{\pi,i} \in K. \tag{8.2}$$

Similarly, in $P_m(R_2)$,

$$x_{\pi(1)} \cdots x_{\pi(m)} = \sum_{j=1}^{c''} \gamma_{\pi,j} h_j(x_1, \ldots, x_m), \quad \gamma_{\pi,j} \in K. \tag{8.3}$$

The equations (8.2) and (8.3) are polynomial identities respectively for R_1 and R_2, and vanish for any $u_1, \ldots, u_m \in R_1$ and $v_1, \ldots, v_m \in R_2$, respectively. We look for a multilinear polynomial identity of degree m for the tensor product $R = R_1 \otimes R_2$ of the K-algebras R_1 and R_2. We consider the expression

$$f(x_1, \ldots, x_m) = \sum_{\pi \in S_m} \xi_\pi x_{\pi(1)} \cdots x_{\pi(m)} = 0$$

with unknown coefficients ξ_π from K and search for conditions which guarantee that $f = 0$ is a polynomial identity for R. Since $f = 0$ is multilinear, it is sufficient to show that it vanishes for arbitrary

$$u_1 \otimes v_1, u_2 \otimes v_2, \ldots, u_m \otimes v_m, \quad u_1, \ldots, u_m \in R_1, \quad v_1, \ldots, v_n \in R_2.$$

We calculate $f(u_1 \otimes v_1, \ldots, u_m \otimes v_m)$ and obtain

$$f(u_1 \otimes v_1, \ldots, u_m \otimes v_m) = \sum_{\pi \in S_m} \xi_\pi (u_{\pi(1)} \otimes v_{\pi(1)}) \cdots (u_{\pi(m)} \otimes v_{\pi(m)})$$

$$= \sum_{\pi \in S_m} \xi_\pi (u_{\pi(1)} \cdots u_{\pi(m)}) \otimes (v_{\pi(1)} \cdots v_{\pi(m)})$$

$$= \sum_{\pi \in S_m} \sum_{i=1}^{c'} \sum_{j=1}^{c''} \xi_\pi \beta_{\pi,i} \gamma_{\pi,j} g_i(u_1, \ldots, u_m) \otimes h_j(v_1, \ldots, v_m) = 0.$$

We rewrite this equation in the form

$$\sum_{i=1}^{c'} \sum_{j=1}^{c''} \left(\sum_{\pi \in S_m} \beta_{\pi,i} \gamma_{\pi,j} \xi_\pi \right) g_i(u_1, \ldots, u_m) \otimes h_j(v_1, \ldots, v_m) = 0.$$

We consider the linear homogeneous system

$$\sum_{\pi \in S_m} \beta_{\pi,i} \gamma_{\pi,j} \xi_\pi = 0, \quad i = 1, \ldots, c', j = 1, \ldots, c''. \tag{8.4}$$

The number of the unknowns ξ_π is $m!$ and the number of the equations is

$$c'c'' = c_m(R_1) c_m(R_2) < m!.$$

Hence the system (8.4) has a nonzero solution ξ_π, $\pi \in S_m$. Clearly, any nontrivial solution of the system corresponds to a nontrivial polynomial identity for $R = R_1 \otimes R_2$. □

Amitsur [A2] proved that every PI-algebra satisfies some power $s_k^d = 0$ of the standard identity $s_k = 0$. His proof uses methods of structure ring theory and does not provide estimates for the values of k and d, see also Chapter 7 of the lecture notes of Formanek in the present book. Here we give an idea of the quantitative proof of Regev [Re2].

Theorem 8.1.7. (Amitsur [A2]) *Every PI-algebra satisfies the polynomial identity*

$$s_k^d(x_1, \ldots, x_k) = \left(\sum_{\sigma \in S_k} (\text{sign } \sigma) x_{\sigma(1)} \cdots x_{\sigma(k)} \right)^d = 0$$

for some $k, d \geq 1$.

Proof. We sketch a simplified version of the proof of Regev [Re2]. (The bounds in [Re2] are better than ours but the proof is more complicated.) We assume that the base field is of characteristic 0 and the PI-algebra R satisfies a polynomial identity of degree n.

The first observation is that the linearization of the polynomial $s_k^d(x_1, \ldots, x_k)$ generates an irreducible S_{kd}-module $M(\lambda)$, where $\lambda = (d^k)$, i.e., the diagram of λ is a rectangle with k rows and d columns. In order to complete the proof it is sufficient to find positive integers k and d such that all irreducible submodules $M(d^k)$ of P_{kd} belong to $T(R)$.

In the second step of the proof we consider the decomposition of the character of the S_m-module P_m (m any):

$$\chi_{S_m}(P_m) = \chi_{S_m}(P_m \cap T(R)) + \chi_{S_m}(P_m/(P_m \cap T(R))),$$

where

$$\chi_{S_m}(P_m) = \sum_{\mu \vdash m} d_\mu \chi_\mu, \quad d_\mu = \dim M(\mu),$$

$$\chi_{S_m}(P_m \cap T(R)) = \sum_{\mu \vdash m} p_\mu \chi_\mu, \quad \chi_{S_m}(P_m/(P_m \cap T(R))) = \chi_m(R) = \sum_{\mu \vdash m} m_\mu \chi_\mu.$$

Hence, if $m_\mu = 0$ for some $\mu \vdash m$, then all irreducible components $M(\mu)$ of P_m belong to $T(R)$. For each $m_\mu \neq 0$ we have that $P_m/(P_m \cap T(R))$ contains a submodule isomorphic to $M(\mu)$. This implies that

$$d_\mu = \dim M(\mu) \leq \dim (P_m/(P_m \cap T(R))) = c_m(R).$$

We shall find k and d such that $\dim M(d^k) > c_{kd}(R)$.

We fix $d > 2(n-1)^2$ and assume that $k \geq d$. Using the hook formula, see Theorem 2.2.5 (iii), we obtain

$$d_{(d^k)} = (kd)! \frac{(d-1)!(d-2)! \cdots 1!0!}{(k+d-1)!(k+d-2)! \cdots (k+1)!k!}$$

which gives the inequality

$$d_{(d^k)} > \frac{(kd)!}{((k+d)!)^d}.$$

Applying the Stirling formula for $m!$

$$m! = \sqrt{2\pi m} \, m^m e^{-m} e^{\theta(m)}, \quad |\theta(m)| < \frac{1}{12m},$$

and, for sufficiently large m,

$$m! \approx \sqrt{2\pi m} m^m e^{-m},$$

we obtain for k sufficiently large

$$d_{(d^k)} > \frac{a}{k^{2d^2}} \left(\frac{d}{2}\right)^{kd}$$

for some positive constant a. Since we have fixed $d > 2(n-1)$, when $k \to \infty$, the fraction

$$\left(\frac{d}{2(n-1)^2}\right)^{kd}$$

goes to infinity faster than the polynomial function k^{2d^2} in $k = (kd)/d$. By the codimension theorem of Regev (Theorem 8.1.5), the codimension sequence of R satisfies the inequality

$$c_m(R) \le (n-1)^{2m}, \quad m = 0, 1, 2, \ldots.$$

Hence for our fixed $d > 2(n-1)^2$ and for k sufficiently large

$$d_{(d^k)} > c_{kd}(R),$$

and this means that R satisfies the identity $s_k(x_1, \ldots, x_k)^d = 0$.   □

8.2   The Theory of Kemer and Growth of Codimensions

We give a very brief account of the structure theory of T-ideals developed by Kemer [Ke2], see also his book [Ke5], in the spirit of ideal theory of polynomial algebras in several variables. We assume that the field K is of characteristic 0. The next definition is similar to the usual definition of semiprime and prime ideals, but instead of ordinary ideals, we consider T-ideals only.

Definition 8.2.1. The T-ideal S of $K\langle X \rangle$ is called *T-semiprime* if any T-ideal U such that $U^k \subseteq S$ for some k, is included in S, i.e., $U \subseteq S$. The T-ideal P is *T-prime* if the inclusion $U_1 U_2 \subseteq P$ for some T-ideals U_1 and U_2 implies $U_1 \subseteq P$ or $U_2 \subseteq P$.

The algebra S is *$\mathbb{Z}_2$-graded* if it is a sum of two subspaces S_0 and S_1 such that $S_i S_j \subseteq S_{i+j}$, where $i, j = 0, 1$, and the subscript $i + j$ is taken modulo 2. Usually one assumes that S is a *direct* sum of S_0 and S_1. In his theory, Kemer does not require $S_0 \cap S_1 = 0$. The Grassmann algebra E of an infinite dimensional vector space with basis $\{e_1, e_2, \ldots\}$ has a natural $\mathbb{Z}_2$-grading $E = E_0 \oplus E_1$, where E_0 and E_1 are spanned on the basis elements $e_{i_1} \cdots e_{i_n}$ of even and odd length, respectively. Let p, q be positive integers, $p \geq q$, and let $M_{p \times q}(E_1)$ be the vector space of all $p \times q$ matrices with entries from E_1. The vector subspace of $M_{p+q}(E)$

$$M_{p,q} = \left\{ \begin{pmatrix} a & b \\ c & d \end{pmatrix} \mid a \in M_p(E_0), b \in M_{p \times q}(E_1), c \in M_{q \times p}(E_1), d \in M_q(E_0) \right\}$$

is an algebra. The building blocks in the theory of Kemer are the polynomial identities of the matrix algebras, the Grassmann algebra and the algebras $M_{p,q}$.

Theorem 8.2.2. (Kemer [Ke2])

(i) *For every T-ideal U of $K\langle X \rangle$ there exists a T-semiprime T-ideal S and a positive integer k such that*

$$S \supseteq U \supseteq S^k.$$

(ii) *Every T-semiprime T-ideal S is an intersection of a finite number of T-prime T-ideals $Q_1, \ldots, Q_m$,*

$$S = Q_1 \cap \cdots \cap Q_m.$$

(iii) *A T-ideal P is T-prime if and only if P coincides with one of the following T-ideals:*

$$T(M_n(K)), T(M_n(E)), T(M_{p,q}), (0), K\langle X \rangle,$$

where $M_n(E)$ is the $n \times n$ matrix algebra with entries from the Grassmann algebra.

The following theorem generalizes the theorem of Kemer (Theorem 7.1.19) that for any finitely generated algebra there exists a finite dimensional algebra with the same polynomial identities. If $S = S_0 + S_1$ is a $\mathbb{Z}_2$-graded algebra, then the algebra

$$E(S) = (E_0 \otimes S_0) \oplus (E_1 \otimes S_1)$$

is called the *Grassmann hull* of S.

Theorem 8.2.3. (Kemer, [Ke2, Ke4], Berele [Be2]) *For any PI-algebra R there exists a finite dimensional $\mathbb{Z}_2$-graded algebra $S = S_0 + S_1$ such that R and the Grassmann hull $E(S)$ of S have the same polynomial identities.*

Kemer [Ke2], Theorem 1, proved that $T(R) = T(S)$ for some finitely generated $\mathbb{Z}_2$-graded algebra $S = S_0 + S_1$. Berele [Be2], Theorem 7, gave another proof with $S = S_0 \oplus S_1$, also finitely generated. The final version of Theorem 8.2.3 was proved by Kemer in [Ke4], Theorem 2. The theorem has the following interesting corollary.

Corollary 8.2.4. *For every PI-algebra R there exists a positive integer n such that $T(R) \supseteq T(M_n(E))$, i.e., R satisfies all polynomial identities of the $n \times n$ matrix algebra $M_n(E)$ with entries from the Grassmann algebra.*

By the codimension theorem of Regev (Theorem 8.1.5), for any PI-algebra R, the sequence $\sqrt[m]{c_m(R)}$, $m = 0, 1, 2, \ldots$, is bounded and hence

$$\limsup_{m \to \infty} \sqrt[m]{c_m(R)} < \infty.$$

There was a conjecture, than $\lim_{m \to \infty} \sqrt[m]{c_m(R)}$ also exists and always is an integer. This was confirmed by Giambruno and Zaicev [GZ1, GZ2]:

Theorem 8.2.5. (Giambruno and Zaicev [GZ1, GZ2]) *For any PI-algebra R the limit*

$$\exp(R) = \lim_{m \to \infty} \sqrt[m]{c_m(R)}$$

exists and is an integer called the exponent of R and of the variety of algebras var R *generated by R.*

The proof of the theorem uses the theory of Kemer combined with combinatorial methods and representation theory of symmetric groups. In particular, Giambruno and Zaicev relate the exponent of the algebra with Theorem 8.2.3 in the following way. We shall state the result in the case of finitely generated PI-algebras only, as in [GZ1]. For the general case of an arbitrary PI-algebra see [GZ2]. By Theorem 7.1.19, if R is a finitely generated PI-algebra, then its T-ideal coincides with the T-ideal of some finite dimensional algebra. Since the extension of the field preserves the polynomial identities, see Remark 1.2.9, we assume that dim $R < \infty$ and the field K is algebraically closed. Let the Wedderburn–Malcev decomposition of the algebra R be

$$R = (B_1 \oplus \cdots \oplus B_k) + J,$$

$B_1 \oplus \cdots \oplus B_k = M_{n_1}(K) \oplus \cdots \oplus M_{n_k}(K)$ is the semisimple component of R and J is the Jacobson radical. Consider all possible ordered s-tuples $(B_{i_1}, \ldots, B_{i_s})$ of pairwise different algebras B_i such that

$$B_{i_1} J B_{i_2} \cdots B_{i_{s-1}} J B_s \neq 0, \quad s = 0, 1, 2, \ldots, k.$$

Then

$$\exp(R) = \max\{\dim B_{i_1} + \cdots + \dim B_{i_s}\}.$$

The exact values of the codimensions are known for very few algebras. Among them are the field $(c_m(K) = 1)$, the Grassmann algebra $(c_m(E) = 2^{m-1}$, $m = 1, 2, \ldots)$, the 2×2 matrix algebra $M_2(K)$, and the tensor product $E \otimes E$ of two copies of the Grassmann algebra. For most of the interesting algebras one knows sufficiently well the asymptotic behaviour of the codimensions only. Of course, using Theorem 2.1.9 one can find exact formulas and asymptotic estimations for the codimensions of products of T-ideals in terms of the codimensions of the factors, as in Corollary 2.1.10.

The standard way to study the codimensions is to obtain information for the cocharacter sequence of the algebra and then to estimate the degree of the corresponding character. The following theorem of Berele and Regev [BR1] is useful for this purpose.

Theorem 8.2.6. *The multiplicities m_λ, $\lambda \vdash m$, in the cocharacter sequence of a PI-algebra R are bounded by a polynomial in m.*

Combining this result with the theorem of Amitsur and Regev (Theorem 2.3.9), one obtains

Corollary 8.2.7. *For every PI-algebra R, there exist integers k, l such that*

$$c_m(R) < f(m)C_{k,l}(m), \quad C_{k,l}(m) = \sum \dim M(\lambda),$$

where $f(m)$ is a polynomial in m and the summation in $C_{k,l}(m)$ is on the partitions $\lambda = (\lambda_1, \ldots, \lambda_p)$ of m satisfing the condition $\lambda_{k+1} \le l$.

For example, easy estimates show that

$$C_{k,l}(m) \le am^b(k+l)^m$$

for some constants a, b and this gives an upper bound for the exponent of R.

In a series of papers (see [Re5] for a survey) Regev obtained the asymptotic behaviour of the codimensions $c_m(M_2(K))$ and the cocharacters $\chi_m(M_2(K))$; the explicit expression of $\chi_m(M_2(K))$ was found by Formanek [F3] and the author [Dr3]; Procesi [Pr4] computed the codimension sequence of $M_2(K)$.

Theorem 8.2.8. *The codimension series and the codimension sequence of $M_2(K)$ are the following:*

$$c(M_2(K), t) = \sum_{m \ge 0} c_m(M_2(K))t^m$$

$$= \frac{1}{t^2}\left(1 - 2t - \sqrt{1-4t}\right) - \frac{t^3}{(1-t)^4} + \frac{1}{1-t} - \frac{1}{1-2t};$$

$$c_m(M_2(K)) = \frac{1}{m+2}\binom{2m+2}{m+1} - \binom{m}{3} + 1 - 2^m.$$

The codimensions $c_m(M_2(K))$ satisfy the asymptotic equality

$$c_m(M_2(K)) \approx \frac{4^{m+1}}{m\sqrt{\pi m}}.$$

Popov [Po] found that the so called proper cocharacter sequence $\xi_m(E \otimes E)$ of $E \otimes E$ has the form

$$\xi_m(E \otimes E) = \sum_{p+2q+r=m} \chi_{(p,2^q,1^r)}, \quad (p, 2^q, 1^r) \neq (m), (1^{2n+1}), \quad m, n > 0,$$

which allows to find the explicit form of the cocharacters, see, e.g., Carini and Di Vincenzo [CDV]. Later, the author calculated the codimensions:

Theorem 8.2.9. (Drensky, [Dr6])

$$c(E \otimes E, t) = \sum_{m \geq 0} c_m(E \otimes E)t^m = \frac{1}{2} + \frac{1}{2\sqrt{1-4t}} + \frac{t}{1-t^2} + \frac{1}{1-t} - \frac{1}{1-2t}.$$

$$c_m(E \otimes E) = \frac{1}{2}\binom{2m}{m} + m + 1 - 2^m, \quad m > 0,$$

$$c_m(E \otimes E) \approx \frac{4^m}{2\sqrt{\pi m}}.$$

Using his method, Regev (see [Re5]) determined the asymptotic behaviour of the codimension sequence of $M_n(K)$ for any $n > 1$ and showed that

$$c_m(M_n(K)) \approx (2\pi)^{(1-n)/2}2^{(1-n^2)/2}1!2!\cdots(n-1)!n^{(n^2+4)/2}m^{(1-n^2)/2}n^{2m+2}.$$

In all cases when the exact asymptotics of $c_m(R)$ is known, it is either of polynomial growth or of the form

$$c_m(R) \sim \alpha m^g d^m, \tag{8.5}$$

where $d \geq 2$ is an integer, g is a half-integer and α is a constant. Regev conjectured that a relation of type (8.5) holds for any PI-algebra. For the moment for the other algebras with T-prime T-ideals the estimates of the codimensions are of the form

$$\alpha_1 m^{g_1} d^m \leq c_m(R) \leq \alpha_2 m^{g_2} d^m,$$

with possibly different α_1, α_2 and g_1, g_2, but with the same d. We shall give only the exponents of $M_n(E)$ and $M_{k,l}$:

$$\exp(M_n(E)) = 2n^2, \quad \exp(M_{k,l}) = k^2 + l^2.$$

For more details see the survey article by Regev [Re7].

The exponent of a PI-algebra can serve as a scale for the complexity of the polynomial identities. If $T(R_1) \subset T(R_2)$, then R_2 has more identities than R_1 and $\exp(R_1) \geq \exp(R_2)$. It is interesting to describe the T-ideals $T(R)$ which are "essentially more complicated" than all T-ideals $T(R_1)$ properly containing $T(R)$.

Definition 8.2.10. (Giambruno and Zaicev [GZ3]) Let $\mathfrak{V}$ be a variety of algebras of exponent e. It is called *minimal of exponent e* if the exponent of every proper subvariety $\mathfrak{W}$ of $\mathfrak{V}$ is smaller than e.

In 1987 the author [Dr5] conjectured that all minimal varieties correspond to products of T-prime T-ideals. (In [Dr5] the minimal varieties were called *extremal* with respect to the codimensions.)

Conjecture 8.2.11. *The variety $\mathfrak{V}$ is minimal of some exponent e if and only if*

$$T(\mathfrak{V}) = T(\mathfrak{S}_1) \cdots T(\mathfrak{S}_k),$$

where $T(\mathfrak{S}_i)$ are T-prime T-ideals, i.e., the varieties $\mathfrak{S}_i$ are generated by some of the algebras $M_n(K)$, $M_n(E)$ or $M_{p,q}$.

The paper [Dr5] contained partial results and discussions confirming this conjecture. The complete answer has been recently obtained by Giambruno and Zaicev, see [GZ3] for the case of varieties generated by finitely generated algebras (equivalently, by finite dimensional algebras) and [GZ4] for the general case.

Theorem 8.2.12. *The variety $\mathfrak{V}$ is minimal of exponent e if and only if $T(\mathfrak{V})$ is a product of T-prime T-ideals $P_1 \cdots P_k$. Then $\exp(\mathfrak{V})$ is equal to the sum of the exponents of the codimension sequences of $P_1, \ldots, P_k$.*

We shall complete this section with a theorem of Leron and Vapne [LV] on the tensor product of algebras with the same polynomial identities and some related topics.

Definition 8.2.13. Two algebras R_1 and R_2 are *PI-equivalent* (notation $R_1 \equiv R_2$) if they have the same polynomial identities, i.e., $T(R_1) = T(R_2)$ (or, equivalently, R_1 and R_2 generate the same variety of algebras).

Before to state the next lemma, we want to mention the abstract characterization of varieties of arbitrary algebraic systems given by Birkhoff.

Theorem 8.2.14. (Birkhoff) *A class of algebras $\mathfrak{V}$ is a variety if and only if $\mathfrak{V}$ is closed under taking Cartesian sums (i.e., "direct sums" allowing infinite support), subalgebras and factor algebras.*

It follows from the proof of the theorem of Birkhoff, that if R is a PI-algebra, then

$$\mathfrak{V} = \text{var } R = \mathcal{QSC}(R).$$

Here we denote by $\mathcal{C}\mathfrak{W}$, $\mathcal{S}\mathfrak{W}$ and $\mathcal{Q}\mathfrak{W}$, respectively, the classes obtained from the class $\mathfrak{W}$ by taking all Cartesian sums, subalgebras and factor algebras of algebras in $\mathfrak{W}$. The proof gives that the relatively free algebras of $\mathfrak{V}$ are subalgebras in Cartesian powers of R (and then every algebra from $\mathfrak{V}$ is a homomorphic image of some relatively free algebra). We shall include the proof of the part which we need for the theorem of Leron and Vapne.

Lemma 8.2.15. (See, e.g., Leron and Vapne [LV], Lemma 2.)

(i) *Let R be a PI-algebra generated by a set of some cardinality $\aleph$. Then there exists a set of indices I such that the relatively free algebra $F_\aleph(\text{var } R)$ of the same cardinality $\aleph$ is embedded into the Cartesian power R^I and R is a homomorphic image of $F_\aleph(\text{var } R)$.*

(ii) *If R_1 and R_2 are algebras such that R_2 is embedded in some Cartesian power R_1^I of R_1 and R_1 is a homomorphic image of R_2, then $R_1 \equiv R_2$.*

Proof. The only nontrivial part of the proof is that $F_\aleph(\mathrm{var}\ R)$ can be embedded into R^I. Let $A = A_\aleph = K\langle X_\aleph\rangle$ be the free algebra of rank $\aleph$. We consider the set $T = T_\aleph(R) \subset A_\aleph$ of all polynomial identities for R. (Usually we take the polynomial identities from the free algebra of countable rank, but this is not essential.) We fix $I = A - T$. Then R^I consists of "$|I|$-tuples" $y = (r_f \mid r_f \in R, f \in I)$. If $f = f(x_{a_1}, \ldots, x_{a_m}) \in I$, this means that $f = 0$ is not a polynomial identity for R and there exist $r_{a_1,f}, \ldots, r_{a_m,f} \in R$ such that $f(r_{a_1,f}, \ldots, r_{a_m,f}) \neq 0$. We define a subset $Y_\aleph$ of R^I as follows. The f-th coordinate of $y_b \in Y_\aleph$ is equal to $r_{a_j,f}$ if b is some of the indices $a_1, \ldots, a_m$. Otherwise the coordinate is set equal to 0. There is a natural homomorphism $\pi : A_\aleph \to R^I$ sending x_b to y_b, $x_b \in X_\aleph$. Clearly, the polynomial identities of R vanish on R^I and $T_\aleph \subseteq \ker \pi$. If $g = g(x_{a_1}, \ldots, x_{a_m})$ does not belong to $T_\aleph(R)$, then $g \in I$ and $g(r_{a_1,g}, \ldots, r_{a_m,g}) \neq 0$. Hence the g-th coordiante of $g(y_{a_1}, \ldots, y_{a_m})$ is different from 0 and g does not belong to $\ker \pi$. In this way $\ker \pi = T_\aleph$ and the subalgebra of R^I generated by $Y_\aleph$ is isomorphic to $A_\aleph/T_\aleph(R)$ which is the relatively free algebra of rank $\aleph$ in $\mathrm{var}\ R$.   □

Theorem 8.2.16. (Leron and Vapne [LV]) *If $R_1 \equiv R_2$ and $S_1 \equiv S_2$ are pairs of PI-equivalent algebras, then the tensor products $R_1 \otimes S_1$ and $R_2 \otimes S_2$ are also PI-equivalent.*

Proof. Since R_1 and R_2 are PI-equivalent, they generate the same variety of algebras $\mathfrak{V}$. Extending the sets of generators of R_1 and R_2 with zeros, we may assume that both algebras have generating sets of the same cardinality $\aleph$. We consider the relatively free algebra $F_\aleph(\mathfrak{V})$ of rank $\aleph$. By Lemma 8.2.15 there exists a set of indices I such that $F_\aleph(\mathfrak{V})$ is embedded into R_1^I and R_1 is a homomorphic image of $F_\aleph(\mathfrak{V})$. We shall denote this as $R_1^I \supset F_\aleph(\mathfrak{V}) \to R_1$. Tensoring with S_1 we obtain

$$R_1^I \otimes S_1 \supset F_\aleph(\mathfrak{V}) \otimes S_1 \to R_1 \otimes S_1.$$

Embedding $R_1^I \otimes S_1$ into $(R_1 \otimes S_1)^I$, we derive

$$(R_1 \otimes S_1)^I \supset F_\aleph(\mathfrak{V}) \otimes S_1 \to R_1 \otimes S_1.$$

By Lemma 8.2.15 again, we obtain that $R_1 \otimes S_1$ has the same polynomial identities as $F_\aleph(\mathfrak{V}) \otimes S_1$, i.e., $R_1 \otimes S_1 \equiv F_\aleph(\mathfrak{V}) \otimes S_1$. Similarly $R_2 \otimes S_1 \equiv F_\aleph(\mathfrak{V}) \otimes S_1$ and this gives that $R_1 \otimes S_1 \equiv R_2 \otimes S_1$. In the same way $R_2 \otimes S_1 \equiv R_2 \otimes S_2$ and, therefore, $R_1 \otimes S_1 \equiv R_2 \otimes S_2$.   □

Corollary 8.2.17. (Leron and Vapne [LV]) *If $R_1 \equiv R_2$, then $M_n(R_1) \equiv M_n(R_2)$.*

Proof. The corollary follows immediately from Theorem 8.2.16, because the PI-equivalence of R_1 and R_2 gives the PI-equivalence of $R_1 \otimes M_n(K)$ and $R_2 \otimes M_n(K)$ which are isomorphic to $M_n(R_1)$ and $M_n(R_2)$, respectively.   □

Theorem 8.2.16 gives that the PI-equivalence is a congruence relative to the tensor product, and therefore the classes of PI-equivalent algebras form a commutative semigroup with unity. It follows from the theory of Kemer that the

PI-equivalence classes of algebras with T-prime T-ideals form a subsemigroup. Below we give the multiplication table for the tensor product of these algebras:

$$M_{k,l} \otimes M_{p,q} \equiv M_{r,s}, \quad r = kp + lq, s = kq + lp,$$

$$M_{p,q} \otimes E \equiv M_{p+q}(E),$$

$$E \otimes E \equiv M_{1,1},$$

together with the trivial isomorphisms

$$M_p(K) \otimes M_q(K) \cong M_{pq}(K), \quad E \otimes M_n(K) \cong M_n(E).$$

Regev [Re6] developed elementary tools in order to give a direct proof of the multiplication rules of Kemer. Another proof involving trace identities of T-prime algebras was given by Berele [Be5]. A new proof of the equivalence $E \otimes E \equiv M_{1,1}$ was obtained by Di Vincenzo [DV3] as a result of the description of the $\mathbb{Z}_2$-graded polynomial identities of $M_{1,1}$. Finally, Di Vincenzo and Nardozza [DVN] derived the equivalence $M_{p,q} \otimes E \equiv M_{p+q}(E)$ using $\mathbb{Z}_{p+q} \times \mathbb{Z}_2$-graded identities.

Additional information about the polynomial identities and the cocharacter sequence of the tensor product of two PI-algebras over a field of characteristic 0 can be found in the paper of Regev [Re4], and for generalizations in positive characteristic in the paper by Di Vincenzo [DV1].

Bibliography

[AP]   S. Abeasis, M. Pittaluga, *On a minimal set of generators for the invariants of* 3×3 *matrices*, Comm. Algebra **17** (1989), 487–499.

[ALB]   J. Alev, L. Le Bruyn, *Automorphisms of generic 2 by 2 matrices*, in "Perspectives in Ring Theory" (Antwerp, 1987), NATO Adv. Sci. Inst. Ser. C: Math. Phys. Sci. **233**, Kluwer Acad. Publ., Dordrecht, 1988, 69–83.

[A1]   S.A. Amitsur, *The T-ideals of the free ring*, J. London Math. Soc. **30** (1955), 470–475.

[A2]   S.A. Amitsur, *A note on PI-rings*, Israel J. Math. **10** (1971), 210–211.

[AL]   S.A. Amitsur, J. Levitzki, *Minimal identities for algebras*, Proc. Amer. Math. Soc. **1** (1950), 449–463.

[AR]   S.A. Amitsur, A. Regev, *PI-algebras and their cocharacters*, J. Algebra **78** (1982), 248–254.

[AS]   S.A. Amitsur, L.W. Small, *Affine algebras with polynomial identities*, Rend. Circ. Mat. Palermo (2) Suppl. **31** (1993), 9–43.

[ATZ]   H. Aslaksen, Eng-Chye Tan, Chen-Bo Zhu, *Generators and relations of invariants of* 2×2 *matrices*, Comm. Algebra **22** (1994), 1821–1832.

[B1]   A.Ya. Belov, *On a Shirshov basis of relatively free algebras of complexity* n (Russian), Mat. Sb. **135** (1988), 373–384. Translation: Math. USSR Sb. **63** (1988), 363–374.

[B2]   A.Ya. Belov, *About height theorem*, Comm. Algebra **23** (1995), 3551–3553.

[B3]   A.Ya. Belov, *Rationality of Hilbert series of relatively free algebras* (Russian), Uspekhi Mat. Nauk **52** (1997), No. 2, 153–154. Translation: Russian Math. Surveys **52** (1997), 394–395.

[B4]   A.Ya. Belov, *On non-Specht varieties* (Russian), Fundam. Prikl. Mat. **5** (1999), No. 1, 47–66.

[B5]   A.Ya. Belov, *Counterexamples to the Specht problem* (Russian), Mat. Sb. **191** (2000), No. 3, 13–24. Translation: Sb. Math. **191** (2000), No. 3, 329–340.

[BBL]   A.Ya. Belov, V.V. Borisenko, V.N. Latyshev, *Monomial algebras*, Algebra, 4, J. Math. Sci. (New York) **87** (1997), No. 3, 3463–3575.

[BDDK]   F. Benanti, J. Demmel, V. Drensky, P. Koev, *Computational approach to polynomial identities of matrices – a survey*, in "Ring Theory: Polynomial

Identities and Combinatorial Methods, Proc. of the Conf. in Pantelleria";
Eds. A. Giambruno, A. Regev, and M. Zaicev, Lect. Notes in Pure and
Appl. Math. **235**, Dekker, 2003, 141–178.

[BD]   F. Benanti, V. Drensky, *On the consequences of the standard polynomial*,
         Comm. Algebra **26** (1998), 4243–4275.

[BZ]   I.I. Benediktovich, A.E. Zalesskij, *Almost standard identities of matrix
         algebras* (Russian), Dokl. Akad. Nauk BSSR **23** (1979), 201–204.

[Be1]   A. Berele, *Homogeneous polynomial identities*, Israel J. Math. **42** (1982),
         258–272.

[Be2]   A. Berele, *Magnum P.I*, Israel J. Math. **51** (1985), 13–19.

[Be3]   A. Berele, *Generic verbally prime PI-algebras and their GK-dimensions*,
         Comm. Algebra **21** (1993), 1487–1504.

[Be4]   A. Berele, *Rates and growth of PI-algebras*, Proc. Amer. Math. Soc. **120**
         (1994), 1047–1048.

[Be5]   A. Berele, *Supertraces and matrices over Grassmann algebras*, Adv.
         Math. **108** (1994), 77–90.

[BR1]   A. Berele, A. Regev, *Applications of hook Young diagrams to P.I. alge-
         bras*, J. Algebra **82** (1983), 559–567.

[BR2]   A. Berele, A. Regev, *Codimensions of products and of intersections of
         verbally prime T-ideals*, Israel J. Math. **103** (1998), 17–28.

[Bg1]   G.M. Bergman, *A note on growth functions of algebras and semi-
         groups*, Research Note, Univ. of California, Berkeley, 1978, unpublished
         mimeographed notes.

[Bg2]   G.M. Bergman, *Wild automorphisms of free P.I. algebras, and some new
         identities*, preprint.

[Bo]   S. Bondari, *Constructing the polynomial identities and central identities
         of degree < 9 of 3 × 3 matrices*, Lin. Algebra Appl. **258** (1997), 233–249.

[BK]   W. Borho, H. Kraft, *Über die Gelfand–Kirillov Dimension*, Math. Ann.
         220 (1976), 1–24.

[Br]   A. Braun, *The nilpotency of the radical in a finitely generated PI-ring*,
         J. Algebra **89** (1984), 375–396.

[Bu]   W. Burnside, *On an unsettled question in the theory of discontinuous
         groups*, Quart. J. Math. **33** (1902), 230–238.

[C]   L. Carini, *The Poincaré series related to the Grassmann algebra*, Lin.
         Multilin. Algebra **27** (1990), 199–205.

[CDV] L. Carini, O. M. Di Vincenzo, *On the multiplicity of the cocharacters of the tensor square of the Grassmann algebra*, Atti dell'Accademia Peloritana dei Pericolanti, Messina, Classe I di Scienze Fis. Mat. e Nat. **69** (1991), 237–246.

[Ch] Q. Chang, *Some consequences of the standard identity*, Proc. Amer. Math. Soc. **104** (1988), 707–710.

[Ck] G.P. Chekanu, *On local finiteness of algebras* (Russian), Mat. Issled. **105** (1988), 153–171.

[Co] P.M. Cohn, *Free Rings and Their Relations*, Second Edition, London Math. Soc. Monographs **19**, Acad. Press, 1985.

[CR] C.W. Curtis, I. Reiner, *Representation Theory of Finite Groups and Associative Algebras*, Interscience, John Willey and Sons, New York-London, 1962.

[DCP] C. De Concini, C. Procesi, *A characteristic free approach to invariant theory*, Adv. Math. **21** (1976), 330–354.

[DC] J.A. Dieudonné, J.B. Carrell, *Invariant Theory, Old and New*, Academic Press, New York – London, 1971.

[Di] R.P. Dilworth, *A decomposition theorem for partially ordered sets*, Ann. of Math. **51** (1950), 161–166.

[DV1] O.M. Di Vincenzo, *On the Kronecker product of S_n-cocharacters for P. I. algebras*, Lin. Multilin. Algebra **23** (1988), 139–143.

[DV2] O.M. Di Vincenzo, *A note on the identities of the Grassmann algebras*, Boll. Un. Mat. Ital. (7) **5**-A (1991), 307–315.

[DV3] O.M. Di Vincenzo, *On the graded identities of $M_{1,1}(E)$*, Israel J. Math. **80** (1992), 323–335.

[DVN] O.M. Di Vincenzo, V. Nardozza, $\mathbb{Z}_{k+l} \times \mathbb{Z}_2$-*graded polynomial identities for $M_{k,l}(E) \otimes E$*, Rend. Semin. Mat. Univ. Padova **108** (2002), 27–39.

[Do1] M. Domokos, *Eulerian polynomial identities and algebras satisfying a standard identity*, J. Algebra **169** (1994), 913–928.

[Do2] M. Domokos, *New identities for 3×3 matrices*, Lin. Multilin. Algebra **38** (1995), 207–213.

[DoKZ] M. Domokos, S.G. Kuzmin, A.N. Zubkov, *Rings of matrix invariants in positive characteristic*, J. Pure Appl. Algebra **176** (2002), 61–80.

[DRS] P. Doubilet, G.C. Rota, J. Stein, *On the foundations of combinatorial theory*, IX. *Combinatorial methods in invariant theory*, Studies in Appl. Math. **53** (1974), 185–216.

122   *Bibliography*

[Dr1]   V. Drensky, *Representations of the symmetric group and varieties of linear algebras* (Russian), Mat. Sb. **115** (1981), 98–115. Translation: Math. USSR Sb. **43** (1981), 85–101.

[Dr2]   V. Drensky, *A minimal basis for the identities of a second-order matrix algebra over a field of characteristic* 0 (Russian), Algebra i Logika **20** (1981), 282–290. Translation: Algebra and Logic **20** (1981), 188–194.

[Dr3]   V. Drensky, *Codimensions of T-ideals and Hilbert series of relatively free algebras*, J. Algebra **91** (1984), 1–17.

[Dr4]   V. Drensky, *On the Hilbert series of relatively free algebras*, Comm. Algebra **12** (1984), 2335–2347.

[Dr5]   V. Drensky, *Extremal varieties of algebras. I, II* (Russian), Serdica **13** (1987), 320–332; **14** (1988), 20–27.

[Dr6]   V. Drensky, *Explicit formulas for the codimensions of some T-ideals* (Russian), Sibirsk. Mat. Zh. **29** (1988) No. 6, 30–36. Translation: Siberian Math. J. **29** (1988), 897–902.

[Dr7]   V. Drensky, *New central polynomials for the matrix algebra*, Israel J. Math. **92** (1995), 235–248.

[Dr8]   V. Drensky, *Commutative and noncommutative invariant theory*, Math. and Education in Math. Proc. of 24th Spring Conf. of the Union of Bulg. Mathematicians, Sofia, 1995, 14–50.

[Dr9]   V. Drensky, *Gelfand-Kirillov dimension of PI-algebras*, in "Methods in Ring Theory, Proc. of the Trento Conf.", Lect. Notes in Pure and Appl. Math. **198**, Dekker, 1998, 97–113.

[Dr10]   V. Drensky, *Free Algebras and PI-Algebras*, Springer-Verlag, Singapore, 2000.

[Dr11]   V. Drensky, *Defining relations for the algebra of invariants of* 2×2 *matrices*, Algebras and Representation Theory **6** (2003), 193–214.

[DK1]   V. Drensky, A. Kasparian, *Polynomial identities of eighth degree for* 3×3 *matrices*, Annuaire de l'Univ. de Sofia, Fac. de Math. et Mecan., Livre 1, Math. **77** (1983), 175–195.

[DK2]   V. Drensky, A. Kasparian, *A new central polynomial for* 3×3 *matrices*, Comm. Algebra **13** (1985), 745–752.

[DKo]   V. Drensky, P. Koshlukov, *Weak polynomial identities for a vector space with a symmetric bilinear form*, Math. and Education in Math. Proc. of 16-th Spring Conf. of the Union of Bulg. Mathematicians, Sofia, Publ. House of the Bulg. Acad. of Sci., 1987, 213–219.

[DPC]   V. Drensky, G.M. Piacentini Cattaneo, *A central polynomial of low degree for* 4×4 *matrices*, J. Algebra **168** (1994), 469–478.

[DR]   V. Drensky, Ts.G. Rashkova, *Weak polynomial identities for the matrix algebras*, Comm. Algebra **21** (1993), 3779–3795.

[Du]   J. Dubnov, *Sur une généralisation de l'équation de Hamilton-Cayley et sur les invariants simultanés de plusieurs affineurs*, Proc. Seminar on Vector and Tensor Analysis, Mechanics Research Inst., Moscow State Univ. **2/3** (1935), 351–367 (see also Zbl. für Math. **12** (1935), p. 176).

[DI]   J. Dubnov, V. Ivanov, *Sur l'abaissement du degré des polynômes en affineurs*, C.R. (Doklady) Acad. Sci. USSR **41** (1943), 96–98 (see also MR **6** (1945), p. 113, Zbl. für Math. **60** (1957), p. 33).

[F1]   E. Formanek, *Central polynomials for matrix rings*, J. Algebra **23** (1972), 129–132.

[F2]   E. Formanek, *The polynomial identities of matrices*, Contemp. Math. **13** (1982), 41–79.

[F3]   E. Formanek, *Invariants and the ring of generic matrices*, J. Algebra **89** (1984), 178–223.

[F4]   E. Formanek, *Noncommutative invariant theory*, Contemp. Math. **43** (1985), 87–119.

[F5]   E. Formanek, *Functional equations for character series associated with $n \times n$ matrices*, Trans. Amer. Math. Soc. **294** (1986), 647–663.

[F6]   E. Formanek, *The Polynomial Identities and Invariants of $n \times n$ Matrices*, CBMS Regional Conf. Series in Math. **78**, Published for the Confer. Board of the Math. Sci. Washington DC, AMS, Providence RI, 1991.

[F7]   E. Formanek, *The ring of generic matrices*, J. Algebra **258** (2002), 310–320.

[Fr]   G. Freundeburg, *A survey of counterexamples to Hilbert's fourteenth problem*, Serdica Math. J. **27** (2001), 171–192.

[GS]   A. Giambruno, S.K. Sehgal, *On a polynomial identity for $n \times n$ matrices*, J. Algebra **126** (1989), 451–453.

[GV]   A. Giambruno, A. Valenti, *Central polynomials and matrix invariants*, Israel J. Math. **96** (1996), 281–297.

[GZ1]   A. Giambruno, M. Zaicev, *On codimension growth of finitely generated associative algebras*, Adv. Math. **140** (1998), 145–155.

[GZ2]   A. Giambruno, M. Zaicev, *Exponential codimension growth of PI algebras: An exact estimate*, Adv. Math. **142** (1999), 221–243.

[GZ3]   A. Giambruno, M. Zaicev, *Minimal varieties of algebras of exponential growth*, Adv. Math. **174** (2003), 310–323.

[GZ4]   A. Giambruno, M. Zaicev, *Codimension growth and minimal superalgebras*, Trans. Amer. Math. Soc. **355** (2003), 5091–5117.

[Go]   E.S. Golod, *On nil-algebras and finitely approximable p-groups* (Russian), Izv. Akad. Nauk SSSR, Ser. Mat. **28** (1964), 273–276.

[GoS]   E.S. Golod, I.R. Shafarevich, *On the class field tower* (Russian), Izv. Akad. Nauk SSSR, Ser. Mat. **28** (1964), 261–272.

[Gr1]   A.V. Grishin, *Examples of T-spaces and T-ideals over a field of characteristic 2 without the finite basis property* (Russian, English summary), Fundam. Prikl. Mat. **5** (1999), No. 1, 101–118.

[Gr2]   A.V. Grishin, *On non-Spechtianness of the variety of associative rings that satisfy the identity* $x^{32} = 0$, Electron. Res. Announc. Am. Math. Soc. **6** (2000), No. 7, 50–51.

[Gr3]   A.V. Grishin, *The variety of assiciative rings, which satisfy the identity* $x^{32} = 0$, *is not Specht*, in Editors: D. Krob et al., "Formal Power Series and Algebraic Combinatorics. Proceedings of the 12th International Conference, FPSAC'00", Moscow, Russia, June 26-30, 2000, Springer-Verlag, Berlin, 2000, 686–691.

[GK1]   C.K. Gupta, A.N. Krasilnikov, *A non-finitely based system of polynomial identities which contains the identity* $x^6 = 0$, Quart. J. Math. **53** (2002), 173–183.

[GK2]   C.K. Gupta, A.N. Krasilnikov, *A simple example of a non-finitely based system of polynomial identities*, Comm. Algebra **30** (2002), 4851–4866.

[Gu]   B.G. Gurevich, *Foundations of the Theory of Algebraic Invariants*, Noordhoff, Groningen, 1964.

[H]   M. Hall, Jr., *Combinatorial Theory*, Blaisdell Publ. Comp., Waltham, Mass.-Toronto-London, 1967. Second Edition, Wiley-Interscience Series in Discrete Math., John Wiley & Sons, New York, 1986.

[Ha1]   P. Halpin, *Central and weak identities for matrices*, Comm. Algebra **11** (1983), 2237–2248.

[Ha2]   P. Halpin, *Some Poincaré series related to identities of* 2×2 *matrices*, Pacific J. Math. **107** (1983), 107–115.

[He]   I.N. Herstein, *Noncommutative Rings*, Carus Math. Monographs **15**, Wiley and Sons, Inc., New York, 1968.

[Hg]   G. Higman, *On a conjecture of Nagata*, Proc. Camb. Philos. Soc. **52** (1956), 1–4.

[Hi1]   D. Hilbert, *Über die Theorie der algebraischen Formen*, Math. Ann. **36** (1890), 473–534; reprinted in "Gesammelte Abhandlungen, Band II, Algebra, Invariantentheorie, Geometrie", Zweite Auflage, Springer-Verlag, Berlin – Heidelberg – New York, 1970, 199–257.

[Hi2]   D. Hilbert, *Mathematische Probleme*, Archiv f. Math. u. Phys. **1** (1901), 44–63, 213–237; reprinted in "Gesammelte Abhandlungen, Band III, Analysis, Grundlagen der Mathematik, Physik, Verschiedenes, Lebensgeschichte", Zweite Auflage, Springer-Verlag, Berlin – Heidelberg – New York, 1970, 290–329.

[J]   N. Jacobson, *PI-Algebras: An Introduction*, Lecture Notes in Math. **441**, Springer-Verlag, Berlin – New York, 1975.

[JK]   G. James, A. Kerber, *The Representation Theory of the Symmetric Group*, Encyclopedia of Math. and Its Appl. **16**, Addison-Wesley, Reading, Mass. 1981.

[Ka1]   I. Kaplansky, *Topological representations of algebras*. I, Trans. Amer. Math. Soc. **68** (1950), 62–75.

[Ka2]   I. Kaplansky, *Problems in the theory of rings*, Report of a Conference on Linear Algebras, June, 1956, in National Acad. of Sci. – National Research Council, Washington, Publ. **502** (1957), 1–3.

[Ka3]   I. Kaplansky, *Problems in the theory of rings revised*, Amer. Math. Monthly **77** (1970), 445–454.

[Ke1]   A.R. Kemer, *Capelli identities and nilpotency of the radical of finitely generated PI-algebras* (Russian), Dokl. Akad. Nauk SSSR **255** (1980), 793–797. Translation: Sov. Math. Dokl. **22** (1980), 750–753.

[Ke2]   A.R. Kemer, *Varieties and $\mathbb{Z}_2$-graded algebras* (Russian), Izv. Akad. Nauk SSSR, Ser. Mat. **48** (1984), 1042–1059. Translation: Math. USSR, Izv. **25** (1985), 359–374.

[Ke3]   A.R. Kemer, *Finite basis property of identities of associative algebras* (Russian), Algebra i Logika **26** (1987), 597–641. Translation: Algebra and Logic **26** (1988), 362–397.

[Ke4]   A.R. Kemer, *Representability of reduced-free algebras* (Russian), Algebra i Logika **27** (1988), 274–294. Translation: Algebra and Logic **27** (1989), 167–184.

[Ke5]   A.R. Kemer, *Ideals of Identities of Associative Algebras*, Translations of Math. Monographs **87**, AMS, Providence, RI, 1991.

[Kh]   V.K. Kharchenko, *A remark on central polynomials* (Russian), Mat. Zametki **26** (1979), 345–346. Translation: Math. Notes **26** (1979), p. 665.

[Ki]   A.A. Kirillov, *Certain division algebras over a field of rational functions* (Russian), Funkts. Anal. Prilozh. **1** (1967), No.1, 101–102. Translation: Funct. Anal. Appl. **1** (1967), 87–88.

[Ko]   P. Koshlukov, *Finitely based ideals of weak polynomial identities*, Comm. Algebra **26** (1998), 3335–3359.

[Ks]   B. Kostant, *A theorem of Frobenius, a theorem of Amitsur–Levitzki and cohomology theory*, J. Math. Mech. **7** (1958), 237–264.

[KR]   D. Krakowski, A. Regev, *The polynomial identities of the Grassmann algebra*, Trans. Amer. Math. Soc. **181** (1973), 429–438.

[KL]   G.R. Krause, T.H. Lenegan, *Growth of Algebras and Gelfand-Kirillov Dimension*, Pitman Publ., London, 1985. Revised edition, Graduate Studies in Mathematics, **22**, AMS, Providence, RI, 2000.

[Kr]   A.G. Kurosch, *Ringtheoretische Probleme, die mit dem Burnsideschen Problem über periodische Gruppen in Zusammenhang stehen* (Russian, German summary), Bull. Acad. Sci. URSS Sér. Math. (Izv. Akad. Nauk SSSR) **5** (1941), 233–240.

[Ku]   E.N. Kuzmin, *On the Nagata-Higman theorem* (Russian), in "Mathematical Structures, Computational Mathematics, Mathematical Modelling. Proc. Dedicated to the 60th Birthday of Acad. L. Iliev", Sofia, 1975, 101–107.

[L]   S. Lang, *Algebra*, Addison-Wesley, Reading, Mass., 1965. Second Edition, 1984.

[La1]   V.N. Latyshev, *Generalization of Hilbert's theorem of the finiteness of bases* (Russian), Sibirsk. Mat. Zh. **7** (1966), 1422–1424. Translation: Sib. Math. J. **7** (1966), 1112–1113.

[La2]   V.N. Latyshev, *On Regev's theorem on identities in a tensor product of PI-algebras* (Russian), Uspekhi Mat. Nauk **27** (1972), No. 4, 213–214.

[LS]   V.N. Latyshev, A.L. Shmelkin, *A certain problem of Kaplansky* (Russian), Algebra i Logika **8** (1969), 447–448. Translation: Algebra and Logic **8** (1969), p. 257.

[LB1]   L. Le Bruyn, *The functional equation for Poincaré series of trace rings of generic 2×2 matrices*, Israel J. Math. **52** (1985), 355–360.

[LB2]   L. Le Bruyn, *Trace Rings of Generic 2 by 2 Matrices*, Memoirs of AMS, **66**, No. 363, Providence, R.I., 1987.

[Le]   U. Leron, *Multilinear identities of the matrix ring*, Trans. Amer. Math. Soc. **183** (1973), 175–202.

[LV]   U. Leron, A. Vapne, *Polynomial identities of related rings*, Israel J. Math. **8** (1970), 127–137.

[Lv]   I. Levitzki, *On a problem of A. Kurosch*, Bull. Amer. Math. Soc. **52** (1946), 1033–1035.

[Lvv]   I.V. Lvov, *On the Shirshov height theorem* (Russian), in "Fifth All-Union Symposium on the Theory of Rings, Algebras and Modules (Held at

Novosibirsk, September, 21st-23rd, 1982)". Abstracts of communications. Institute of Mathematics SO AN SSSR, Novosibirsk 1982, 89–90.

[Mc]   I.G. Macdonald, *Symmetric Functions and Hall Polynomials*, Oxford Univ. Press (Clarendon), Oxford, 1979. Second Edition, 1995.

[Ma]   A.I. Malcev, *On algebras defined by identities* (Russian), Mat. Sb. **26** (1950), 19–33.

[Mts]   Yu.N. Maltsev, *A basis for the identities of the algebra of upper triangular matrices* (Russian), Algebra i Logika **10** (1971), 393–400. Translation: Algebra and Logic **10** (1971), 242–247.

[Mr]   V.T. Markov, *The Gelfand-Kirillov dimension: nilpotency, representability, non-matrix varieties* (Russian), Siberian School on Varieties of Algebraic Systems, Abstracts, Barnaul, 1988, 43–45. Zbl. 685.00002.

[Mo]   T. Molien, *Über die Invarianten der linearen Substitutionsgruppen*, Sitz. König Preuss. Akad. Wiss. **52** (1897), 1152–1156.

[Na1]   M. Nagata, *On the nilpotency of nil algebras*, J. Math. Soc. Japan **4** (1953), 296–301.

[Na2]   M. Nagata, *On the 14th problem of Hilbert*, Amer. J. Math. **81** (1959), 766–772.

[Nk]   K. Nakamoto, *The structure of the invariant ring of two matrices of degree 3*, J. Pure Appl. Algebra **166** (2002), No. 1–2, 125–148.

[Ne]   B.H. Neumann, *Identical relations in groups*. I, Math. Ann. **114** (1937), 506–525.

[No]   E. Noether, *Der Endlichkeitssatz der Invarianten endlicher Gruppen*, Math. Ann. **77** (1916), 89–92; reprinted in "Gesammelte Abhandlungen. Collected Papers", Springer-Verlag, Berlin – Heidelberg – New York – Tokyo, 1983, 181–184.

[Ok1]   S.V. Okhitin, *On varieties defined by two-variable identities* (Russian), Moskov. Gos. Univ. Moscow 1986 (manuscript deposited in VINITI 12.02.1986 No 1016-V), Ref. Zh. Mat. 6A366DEP/1986.

[Ok2]   S.V. Okhitin, *Central polynomials of an algebra of second order matrices* (Russian), Vestnik Moskov. Univ. Ser. I, Mat. Mekh. (1988) No. 4, 61–63. Translation: Moscow Univ. Math. Bull. **43** (1988), No. 4, 49–51.

[OR]   J. Olsson, A. Regev, *On the T-ideal generated by a standard identity*, Israel J. Math. **26** (1977), 97–104.

[Pa]   D.S. Passman, *The Algebraic Structure of Group Rings*, Wiley-Interscience, New York, 1977.

[Pe1]   V.M. Petrogradsky, *On types of super exponential growth of identities in Lie PI-algebras* (Russian), Fundam. Prikl. Mat. **1** (1995), 989–1007.

[Pe2]   V.M. Petrogradsky, *Growth of polynilpotent varieties of Lie algebras and rapidly growing entire functions* (Russian), Mat. Sb. **188** (1997), No. 6, 119–138. Translation: Sb. Math. **188** (1997), 913–931.

[Pe3]   V.M. Petrogradsky, *On the complexity functions for T-ideals of associative algebras* (Russian), Mat. Zametki **68** (2000), 887–897. Translation: Math. Notes **68** (2000), 751–759.

[Po]   A.P. Popov, *Identities of the tensor square of a Grassmann algebra* (Russian), Algebra i Logika **21** (1982), 442–471. Translation: Algebra and Logic **21** (1982), 296–316.

[Pr1]   C. Procesi, *Non-commutative affine rings*, Atti Accad. Naz. Lincei Rend. Cl. Sci. Fis. Mat. Natur. **8** (1967), 239–255.

[Pr2]   C. Procesi, *Rings with Polynomial Identities*, Marcel Dekker, New York, 1973.

[Pr3]   C. Procesi, *The invariant theory of $n \times n$ matrices*, Adv. Math. **19** (1976), 306–381.

[Pr4]   C. Procesi, *Computing with 2×2 matrices*, J. Algebra **87** (1984), 342–359.

[Ra1]   Yu.P. Razmyslov, *Finite basing of the identities of a matrix algebra of second order over a field of characteristic zero* (Russian), Algebra i Logika **12** (1973), 83–113. Translation: Algebra and Logic **12** (1973), 47–63.

[Ra2]   Yu.P. Razmyslov, *On a problem of Kaplansky* (Russian), Izv. Akad. Nauk SSSR, Ser. Mat. **37** (1973), 483–501. Translation: Math. USSR, Izv. **7** (1973), 479–496.

[Ra3]   Yu.P. Razmyslov, *Trace identities of full matrix algebras over a field of characteristic zero* (Russian), Izv. Akad. Nauk SSSR, Ser. Mat. **38** (1974), 723–756. Translation: Math. USSR, Izv. **8** (1974), 727–760.

[Ra4]   Yu.P. Razmyslov, *The Jacobson radical in PI-algebras* (Russian), Algebra i Logika **13** (1974), 337–360. Translation: Algebra and Logic **13** (1974), 192–204.

[Ra5]   Yu.P. Razmyslov, *Identities of Algebras and Their Representations* (Russian), "Sovremennaya Algebra", "Nauka", Moscow, 1989. Translation: Translations of Math. Monographs **138**, AMS, Providence, R.I., 1994.

[Re1]   A. Regev, *Existence of identities in $A \otimes B$*, Israel J. Math. **11** (1972), 131–152.

[Re2]   A. Regev, *The representation of S_n and explicit identities for P.I. algebras*, J. Algebra **51** (1978), 25–40.

[Re3]   A. Regev, *Algebras satisfying a Capelli identity*, Israel J. Math. **33** (1979), 149–154.

[Re4]   A. Regev, *The Kronecker product of S_n-characters and an $A \otimes B$ theorem for Capelli identities*, J. Algebra **66** (1980), 505–510.

[Re5]   A. Regev, On the codimensions of matrix algebras, in "Algebra – Some Current Trends (Varna, 1986)", Lect. Notes in Math. **1352**, Springer-Verlag, Berlin-New York, 162–172, 1988.

[Re6]   A. Regev, *Tensor products of matrix algebras over the Grassmann algebra*, J. Algebra **133** (1990), 512–526.

[Re7]   A. Regev, *Asymptotics of codimensions of some P.I. algebras*, in "Trends in Ring Theory" (Proc. Conf. Miskolc, 1996; Eds. V. Dlab and L. Márki), CMS Conference Proceedings **22**, AMS, Providence, 1998, 159–172.

[Ro]   S. Rosset, *A new proof of the Amitsur–Levitzki identity*, Israel J. Math. **23** (1976), 187–188.

[Rw1]   L.H. Rowen, *Polynomial Identities of Ring Theory*, Acad. Press, 1980.

[Rw2]   L.H. Rowen, *Ring Theory*, vol. 1, 2, Academic Press, Inc., Boston, MA, 1988. Student Edition, Academic Press, Inc., Boston, MA, 1991.

[Shc]   V.V. Shchigolev, *Examples of infinitely based T-ideals* (Russian, English summary), Fundam. Prikl. Mat. **5** (1999), No. 1, 307–312.

[SZ]   I.P. Shestakov, N. Zhukavets, *On associative algebras satisfying the identity $x^5 = 0$*, preprint.

[Sh]   A.I. Shirshov, *On rings with identity relations* (Russian), Mat. Sb. **41** (1957), 277–283.

[Si]   K.S. Sibirskii, *Algebraic invariants for a set of matrices* (Russian), Sib. Mat. Zh. **9** (1968), No. 1, 152–164. Translation: Siber. Math. J. **9** (1968), 115–124.

[Sp]   W. Specht, *Gesetze in Ringen*. I, Math. Z. **52** (1950), 557–589.

[St1]   R.P. Stanley, *Combinatorics and invariant theory*, in "Relations between Combinatorics and Other Parts of Mathematics" Ed. D. K. Ray-Chaudhuri, Proc. Sympos. Pure Math., **34**, AMS, Providence, RI, 1979, 345–355.

[St2]   R.P. Stanley, *Combinatorics and Commutative Algebra*, Progress in Math. **41**, Birkhäuser Boston, Boston, MA, 1983.

[Sw]   R.G. Swan, *An application of graph theory to algebra*, Proc. Amer. Math. Soc. **14** (1963), 367–373. Correction: **21** (1969), 379–380.

[STR]   J. Szigeti, Z. Tuza, G. Révész, *Eulerian polynomial identities on matrix rings*, J. Algebra **161** (1993), 90–101.

[T1]   Y. Teranishi, *The ring of invariants of matrices*, Nagoya Math. J. **104** (1986), 149–161.

[T2]   Y. Teranishi, *Linear diophantine equations and invariant theory of matrices*, "Commut. Algebra and Combinatorics (Kyoto, 1985)", Adv. Stud. Pure Math. **11**, North-Holland, Amsterdam – New York, 1987, 259–275.

[T3]   Y. Teranishi, *The Hilbert series of matrix concominants*, Nagoya Math. J. **111** (1988), 143–156.

[U1]   V.A. Ufnarovski, *An independence theorem and its consequences* (Russian), Mat. Sb. **128** (1985), 124–132. Translation: Math. USSR Sb. **56** (1985), 121–129.

[U2]   V.A. Ufnarovski, *Combinatorial and asymptotic methods in algebra*, in A.I. Kostrikin, I.R. Shafarevich (Eds.), "Algebra VI", Encyclopedia of Math. Sciences **57**, Springer-Verlag, 1995, 1–196.

[VB1]   M. Van den Bergh, *Trace rings are Cohen-Macaulay*, J. Amer. Math. Soc. **2** (1989), 775–799.

[VB2]   M. Van den Bergh, *Explicit rational forms for the Poincaré series of the trace rings of generic matrices*, Israel J. Math. **73** (1991), 17–31.

[VL]   M.R. Vaughan-Lee, *An algorithm for computing graded algebras*, J. Symbolic Comput. **16** (1993), 345–354.

[Vi]   U. Vishne, *Polynomial identities of $M_2(G)$*, Comm. Algebra **30** (2002), 443–454.

[Wa]   W. Wagner, *Über die Grundlagen der projektiven Geometrie und allgemeine Zahlsysteme*, Math. Z. **113** (1937), 528–567.

[We1]   H. Weyl, *Zur Darstellungstheorie und Invariantenabzählung der projektiven, der Komplex- und der Drehungsgruppe*, Acta Math. **48** (1926), 255–278; reprinted in "Gesammelte Abhandlungen", Band III, Springer-Verlag, Berlin – Heidelberg – New York, 1968, 1–25.

[We2]   H. Weyl, *The Classical Groups, Their Invariants and Representations*, Princeton Univ. Press, Princeton, N.J., 1946, New Edition, 1997.

[ZSSS] K.A. Zhevlakov, A.M. Slinko, I.P. Shestakov, A.I. Shirshov, *Rings That Are Nearly Associative* (Russian), "Nauka", Moscow, 1978. Translation: Academic Press, New York, 1982.

Part B

Polynomial Identity Rings

Edward Formanek

Introduction

This second part is an introduction to the structural side of polynomial identity rings (PI-rings).

The introductory Chapter 1 presents definitions and elementary properties of polynomial identities. Chapter 2 proves the Amitsur–Levitzki Theorem and Chapter 3 gives the E. Formanek construction of central polynomials for $n \times n$ matrices. (The Y.P. Razmyslov construction, which was independently discovered at about the same time, is given in Drensky's notes.) Although both results are proved by combinatorial arguments, they are of fundamental importance in both the structural and combinatorial sides of PI-theory.

Chapters 4–9 prove the three major structure theorems of PI-theory, the theorems of I. Kaplansky, E.C. Posner, and M. Artin. Kaplansky's Theorem says that a primitive PI-ring is central simple and finite-dimensional over its center. Kaplansky proved it in a foundational paper [K] which introduced the term *polynomial identity*. His theorem remains the single most important theorem about polynomial identity rings. Posner's Theorem says that a prime PI-ring has a ring of quotients which is a primitive PI-ring. Artin's Theorem characterizes *Azumaya algebras* (also called *central separable algebras*) of constant rank n^2 (as projective modules over their centers) as rings which satisfy the polynomial identities satisfied by $M_n(\mathbb{Z})$ and have no nonzero homomorphic images which satisfy the polynomial identities satisfied by $M_{n-1}(\mathbb{Z})$.

Since Azumaya algebras are finite modules over their centers, the common theme of these theorems is that the PI-hypothesis (in conjunction with a suitable further hypothesis) gives rise to a large center. All three theorems were proved before the construction of central polynomials, but only the proof of Kaplansky's Theorem remains as it was in the original paper. The proofs of both Posner's Theorem and Artin's Theorem have been greatly simplified through the use of central polynomials. In addition, the ring of quotients in Posner's Theorem was originally obtained by noncommutative localization via Goldie's Theorem, while in the modern version it is only necessary to invert the nonzero elements of the center.

An important theme in structural PI-theory is the extension of theorems about commutative rings to PI-rings. Chapters 10 and 11 contain representative theorems of this kind. I did not include the two deepest examples: W. Schelter's theorem that PI-rings which are finitely generated as algebras over a field are *catenary* (maximal chains of prime ideals between two prime ideals have the same length) and the Y.P. Razmyslov–A.R. Kemer–A. Braun theorem that the Jacobson radical is nilpotent in a PI-ring which is finitely generated as an algebra over a commutative Noetherian ring.

The final Chapters 12–16 introduce the ring of $n \times n$ generic matrices over a field K, which is the K-algebra generated by set of (at least two) $n \times n$ *generic matrices*, where an $n \times n$ generic matrix is an $n \times n$ matrix whose entries are independent commuting indeterminates over K. By Posner's Theorem, the ring of generic matrices has a quotient field which is called the *generic division ring*. After establishing some of the properties of the ring of generic matrices, the notes conclude with a a discussion of a major open problem: Is the center of the generic division ring a (stably) pure transcendental extension of K? (A field L containing K is *stably* a pure transcendental extension of K if some pure transcendental extension of L is a pure transcendental extension of K.)

These notes are only an introduction. For readers who wish to learn more, the next page has a list of books, research monographs and survey articles with further material on PI-rings.

General References for PI-Rings

S.A. Amitsur, *Polynomial Identities*, Israel J. Math. **19** (1974), 183–199.

Y. Bahturin, *Basic Structures of Modern Algebra*, Kluwer, 1993.

P.M. Cohn, *Algebra*, Volume 2, Wiley, 1977.

V.S. Drensky, *Free Algebras and PI-Algebras*, Springer-Verlag, 2000.

E. Formanek, *The polynomial identities of matrices*, pp. 41–79 in *Algebraists' Homage* (S.A. Amitsur et al, eds.), Contemporary Math., Vol. 13, Amer. Math. Soc., 1982.

E. Formanek, *The invariants of $n \times n$ matrices*, pp. 18–43 in *Invariant Theory* (S.S. Koh, ed.), Lecture Notes in Math. Vol. 1278, Springer-Verlag, 1987.

E. Formanek, *The Polynomial Identities and Invariants of $n \times n$ Matrices*, CBMS Regional Conf. Series in Math. No. 78, Amer. Math. Soc., 1991.

A.W. Goldie, *Lectures on Quotient Rings and Rings with Polynomial Identities*, Vorlesungen Math. Inst. Giessen, Heft 1, University of Giessen, 1974.

A.W. Goldie, *Rings with Polynomial Identities*, Carleton University Lecture Notes, Ottawa, 1975.

I.N. Herstein, *Noncommutative Rings*, Carus Math. Monographs No. 15, Math. Assoc. Amer., 1968.

I.N. Herstein, *Notes from a Ring Theory Conference*, CBMS Regional Conf. Series in Math. No. 9, Amer. Math. Soc., 1971.

I.N. Herstein, *Rings with Involution*, Chicago Lectures in Math., Univ. of Chicago Press, Chicago, 1976.

N. Jacobson, *PI-Algebras – an Introduction*, Lecture Notes in Math. Vol. 441, Springer-Verlag, 1975.

A.R. Kemer, *Ideals of Identities of Associative Algebras*, Nauka, 1988 (Russian). Translation: Math. Monographs Vol. 87, Amer. Math. Soc., 1991.

L. LeBruyn, *Trace Rings of Generic 2×2 Matrices*, Mem. Amer. Math. Soc. No. 363 (1987).

J.C. McConnell and J.C. Robson, *Noncommutative Noetherian Rings*, Wiley-Interscience, 1987.

B.J. Mueller, *Rings with Polynomial Identities*, Monografias del Instituto de Matematicas No. 4, Universidad Nacional Autonoma de Mexico, Mexico City, 1976.

D.S. Passman, *Infinite Group Rings* Pure and Appl. Math., No. 6, Dekker, 1971.

D.S. Passman, *The Algebraic Structure of Group Rings*, Wiley-Interscience, 1977.

R.S. Pierce, *Associative Algebras*, Graduate Texts in Math. No. 88, Springer-Verlag, 1982.

C. Procesi, *Rings with Polynomial Identities*, Pure and Appl. Math. No. 25, Dekker, 1973.

Y.P. Razmyslov, *Identities of Algebras and their Representations*, Nauka, 1989 (Russian). Translation: Math. Monographs Vol. 138, Amer. Math. Soc., 1994.

G. Renault, *Algèbre Noncommutative*, Gauthier-Villars, 1975.

L.H. Rowen, *Polynomial Identities in Ring Theory*, Academic Press, 1980.

L.H. Rowen, *Ring Theory*, Vols. I, II, Academic Press, 1988.

Chapter 1

Polynomial Identities

Let R be a ring. Unless the contrary is explicitly stated, we assume that R has a unit 1. Ring homomorphisms $R \to S$ are assumed to be unitary, meaning that the unit of R is carried to the unit of S. "Ideal" means two-sided ideal, and "R-module" means unitary left R-module, where an R-module M is unitary if $1m = m$ for all $m \in M$.

If A is a commutative ring, an A-algebra is a ring R together with a homomorphism from A to the center of R. The homomorphism is not required to be one-to-one, so all rings are $\mathbb{Z}$-algebras. In practice, algebras will be over $\mathbb{Z}$ or a field.

Notation 1.1. If A is a commutative ring, $A\langle x_1, \ldots, x_n \rangle$ denotes a free associative algebra over A in n variables, and $A\langle X \rangle$ denotes a free associative algebra over A in countably many variables $x_1, x_2, \ldots$. (Often it will be convenient to use other variables x, y, z, y_i, z_i, etc.)

Suppose that A is a commutative ring, R is an A-algebra, and $r_1, \ldots, r_n \in R$. If $f(x_1, \ldots, x_n) \in A\langle X \rangle$, then $f(r_1, \ldots, r_n) \in R$ denotes the evaluation of f at $r_1, \ldots, r_n$. It is the image of f under any A-algebra homomorphism $A\langle X \rangle \to R$ which sends $x_i \to r_i$, for $i = 1, \ldots, n$.

The free algebra $A\langle X \rangle$ can be graded by giving each variable degree 1, and multigraded by assigning a separate degree function to each variable.

Definition 1.2. A polynomial $f(x_1, \ldots, x_n) \in A\langle X \rangle$ is *homogeneous of degree r* if it is an A-linear combination of monomials of degree r, where each x_s has degree 1. It is *homogeneous of type $(i_1, \ldots, i_n)$* if it is an A-linear combination of monomials, each of which has degree i_s in x_s, for $s = 1, \ldots, n$. It is *linear in the variable x_i* if each monomial in f has degree 1 in x_i. It is *multilinear in $x_{i_1}, \ldots, x_{i_s}$* if it is linear in each of the variables $x_{i_1}, \ldots, x_{i_s}$.

Note that if $f(x_1, \ldots, x_n)$ is a multilinear polynomial in $x_1, \ldots, x_n$, then f can be expressed in the form

$$f(x_1, \ldots, x_n) = \sum \{ a(\sigma) x_{\sigma(1)} \cdots x_{\sigma(n)} \mid \sigma \in S_n,\ a(\sigma) \in A \},$$

where S_n is the symmetric group of permutations of $\{1, \ldots, n\}$. The obvious action of S_n on $\{x_1, \ldots, x_n\}$ extends to an action of S_n on the multilinear polynomials in $x_1, \ldots, x_n$, namely

$$(\pi f)(x_1, \ldots, x_n) = f(x_{\pi(1)}, \ldots, x_{\pi(n)}) = \sum \{ a(\sigma) x_{\pi\sigma(1)} \cdots x_{\pi\sigma(n)} \}.$$

Since the free algebra $A\langle X\rangle$ is the direct sum of its *homogeneous components* of distinct types, every polynomial in $A\langle X\rangle$ is uniquely expressible as a sum of homogeneous polynomials.

Examples 1.3.   1. $x_1^2 x_2 x_3^3 - 3x_2 x_3 x_1 x_3^2 x_1$ (homogeneous of type $(2,1,3)$ and linear in x_2).

2. $x_1 x_2 x_3 - 3x_3 x_1 x_2$ (multilinear in x_1, x_2, x_3).

3. $x_1^2 x_2 + 2x_2 x_3^2 - x_1 x_2 x_1 - x_3 x_2 x_3 = (x_1^2 x_2 - x_1 x_2 x_1) + (2x_2 x_3^2 - x_3 x_2 x_3)$ is a sum of homogeneous polynomials of types $(2,1,0)$ and $(0,1,2)$. Relative to the ordinary grading where each variable has degree 1, it is homogeneous of degree 3.

Definition 1.4. Suppose that A is a commutative ring, R is an A-algebra, and $f(x_1, \ldots, x_n) \in A\langle X\rangle$. Then f is a *polynomial identity* (*PI*) for R if $f(r_1, \ldots, r_n) = 0$ for all $r_1, \ldots, r_n \in R$.

Definition 1.5. A *T-ideal* of $A\langle X\rangle$ is a two-sided ideal of $A\langle X\rangle$ which is closed under A-endomorphisms of $A\langle X\rangle$.

Since A-endomorphisms of $A\langle X\rangle$ are given by specializations $x_i \to y_i$, where $y_i \in A\langle X\rangle$, the next result is easily proved.

Theorem 1.6. *Let R be an A-algebra, and let*

$$T(R) = \{f(x_1, \ldots, x_n) \in A\langle X\rangle \mid f(r_1, \ldots, r_n) = 0 \text{ for all } r_1, \ldots, r_n \in R\}.$$

Then $T(R)$ is a T-ideal of $A\langle X\rangle$.
  Moreover, if J is a T-ideal of $A\langle X\rangle$, then $T(A\langle X\rangle / J) = J$. □

Definition 1.7. If R is an A-algebra, $T(R)$ is called the *T-ideal* of identities of R.

Technically, the notation should include a reference to A, but this will always be clear from the context. T-ideals are an analogue of *fully invariant subgroups* in group theory – these are subgroups of a free group $\mathcal{F}$ on countably many generators which are invariant under endomorphisms of $\mathcal{F}$.

Theorem 1.8. *Suppose that the commutative ring A contains an infinite field K. Then every T-ideal J in $A\langle X\rangle$ is homogeneous. (Every homogeneous component of a polynomial in J lies in J.)*

Proof. If $f(x_1, \ldots, x_n)$ is homogeneous of type $(i_1, \ldots, i_n)$ and $\lambda_1, \ldots, \lambda_n \in K$, then $f(\lambda_1 x_1, \ldots, \lambda_n x_n) = \lambda_1^{i_1} \cdots \lambda_n^{i_n} f(x_1, \ldots, x_n)$. This allows the homogeneous components of a general polynomial g to be expressed as K-linear combinations of specializations of g via a Vandermonde determinant argument. □

Theorem 1.9. *Let R be a K-algebra, where K is an infinite field, and let L be an extension field of K. Then R and $R \otimes_K L$ satisfy the same polynomials in $K\langle X\rangle$.*

Proof. Let $\{r_\alpha\}$ be a basis for R over K, and thus also a basis for $R \otimes_K L$ over L, and let $\{t_{\alpha,i} \mid i = 1, 2, \dots\}$ be a set of independent central variables over R. Suppose that $f(x_1, \dots, x_n) \in K\langle X \rangle$ is a polynomial identity for R. Evaluate f under the substitution $x_i \to t_{\alpha_1,i} r_{\alpha_1} + t_{\alpha_2,i} r_{\alpha_2} + \cdots + t_{\alpha_k,i} r_{\alpha_k}$, where $r_{\alpha_1}, \cdots, r_{\alpha_k}$ are finitely many elements of the basis. The result is equal to a finite sum $\sum F_\beta(t_{\alpha,i}) r_\beta$, where $F_\beta(t_{\alpha,i}) \in K[t_{\alpha,i}]$. The fact that f is a PI for R means that the $F_\beta(t_{\alpha,i})$ vanish under all evaluations in the infinite field K, so they are identically zero. Thus they also vanish under all evaluations in L, which means that f is a PI for $R \otimes_K L$. $\qquad\square$

We state the next theorem without proof. It is a basic fact in combinatorial PI-theory, but is not essential for structural PI-theory.

Theorem 1.10. *Suppose that A is a commutative $\mathbb{Q}$-algebra. Then every T-ideal in $A\langle X \rangle$ is generated as a T-ideal by the multilinear polynomials it contains.* $\qquad\square$

For any A-algebra R, a polynomial $f(x_1, \dots, x_n) \in A\langle X \rangle$ defines a function $R^n \to R$ via specialization. If f is linear in the variable x_i, then this function is A-linear as a function of x_i. This implies the following lemma.

Lemma 1.11. *Let A be a commutative ring, and let $f(x_1, \dots, x_n) \in A\langle X \rangle$ be multilinear in $x_1, \dots, x_n$. Suppose that R is an A-algebra which is generated as an A-module by a set T. If $f(t_1, \dots, t_n) = 0$ for all $t_1, \dots, t_n \in T$, then f is a polynomial identity for R.*

Proof. If $r_1, \dots, r_n \in R$, each r_i is an A-linear combination of elements of T. The multilinearity of f implies that $f(r_1, \dots, r_n)$ is an A-linear combination of evaluations $f(t_1, \dots, t_n)$, with $t_i \in T$. Hence $f(r_1, \dots, r_n) = 0$. $\qquad\square$

Definition 1.12. A polynomial $f(x_1, \dots, x_n) \in A\langle X \rangle$ is *proper* if some coefficient in the highest degree homogeneous component of f (with respect to the ordinary grading) is equal to 1.

Definition 1.13. Let R be an A-algebra, where A is a commutative ring. Then R is a *polynomial identity algebra over A* (or, R is a *PI-algebra over A*, or R *satisfies a PI over A*) if there is a proper polynomial $f \in A\langle X \rangle$ such that f is a polynomial identity for R. If $A = \mathbb{Z}$, then we say that R is a *polynomial identity ring* (or, R is a *PI-ring*, or R *satisfies a PI*).

Some explanation is in order for the technical condition that a coefficient of the highest degree homogeneous component of f must equal 1. Let K be a field of characteristic $p > 0$, and let $R = K\langle X \rangle$. Then only $0 \in K\langle X \rangle$ is a PI for R over K, but $px_1 \in \mathbb{Z}\langle X \rangle$ is a nonzero PI for R over $\mathbb{Z}$. Since we do not want R to be a PI-ring, the condition that some coefficient equals 1 is introduced. The further condition that 1 is a coefficient of the highest degree homogeneous component is needed in order to prove Theorem 1.14 below, which says that a PI-ring always satisfies a multilinear proper PI.

The requirement that some coefficient of f equals 1 avoids the possibility that an A-algebra R might satisfy a PI over $\mathbb{Z}$ but not over A. It does not preclude the possibility that R might satisfy a PI over A but not a PI over $\mathbb{Z}$, which is a deeper question. Later, we will prove a result of Amitsur (Theorem 7.3) which implies that if R satisfies a PI over A, then it satisfies a PI over $\mathbb{Z}$.

The next theorem also belongs to combinatorial PI-theory, and its proof is omitted. It does have an important application in the proof of Kaplansky's Theorem (Theorem 4.13).

Theorem 1.14. *Suppose that a proper polynomial $f(x_1, \ldots, x_n) \in A\langle X \rangle$ is a PI for the A-algebra R. Then there is a proper multilinear polynomial $g \in A\langle X \rangle$, of the same degree as f, which is a PI for R.* $\square$

Corollary 1.15. *If R is a PI-ring, then $R[T]$ is also a PI-ring, where T is a a set of central indeterminates over R.*

Proof. By Theorem 1.14, R satisfies a proper multilinear polynomial $f(x_1, \ldots, x_n)$ $\in \mathbb{Z}\langle X \rangle$. By multilinearity, any evaluation of f on $R[T]$ is a sum of evaluations $f(r_1\mu_1, \ldots, r_n\mu_n)$, where $r_1, \ldots, r_n \in R$, and $\mu_1, \ldots, \mu_n$ are monomials from T (i.e., products of elements of T). Since

$$f(r_1\mu_1, \ldots, r_n\mu_n) = f(r_1, \ldots, r_n)\mu_1 \cdots \mu_n = 0$$

by multilinearity, it follows that f is a PI for $R[T]$. $\square$

Examples, where K is a field 1.16.

1. Any commutative ring satisfies $[x, y]$. ($[x, y] = xy - yx$.)

2. Any Boolean algebra satisfies $x^2 - x$.

3. The finite field with q elements satisfies $x^q - x$.

4. The ring of upper triangular $n \times n$ matrices with entries in K satisfies the polynomial $[x_1, y_1][x_2, y_2] \cdots [x_n, y_n]$.

5. The *exterior or Grassmann algebra* $E(V)$ on an infinite-dimensional vector space V over K satisfies $(xy - yx)z - z(xy - yx)$. (See Chapter 2, Definition 2.1, for the definition of $E(V)$.)

6. The ring of 2×2 matrices over K, $M_2(K)$, satisfies $[[x, y]^2, z]$.

7. Any subring or homomorphic image of a PI-ring is a PI-ring.

8. The free algebra, $K\langle X \rangle$, does not satisfy a PI.

9. The *Weyl algebra*, $K\langle x, y \mid xy - yx = 1 \rangle$ does not satisfy a PI if K has characteristic zero. If the characteristic of K is $p > 0$, it does satisfy a PI.

Note that the polynomials in examples 2 and 3 are PI's whose homogeneous components are not PI's, which shows that the hypothesis in Theorem 1.8 that A contains an infinite field is necessary.

Definition 1.17. The *standard polynomial of degree n* is

$$s_n(x_1, \ldots, x_n) = \sum \{ sign(\sigma) x_{\sigma(1)} \cdots x_{\sigma(n)} \mid \sigma \in S_n \}.$$

Definition 1.18. The polynomial $f(x_1, \ldots, x_n) \in A\langle X \rangle$ is *alternating* in the variables $x_{i_1}, \ldots, x_{i_s}$ if it vanishes under any evaluation in which two of $x_{i_1}, \ldots, x_{i_s}$ are set equal.

The two key properties of the standard polynomial $s_n(x_1, \ldots, x_n)$ are that it is multilinear and alternating in $x_1, \ldots, x_n$. Another important polynomial is the *Capelli polynomial*

$$\begin{aligned} d_n(x_1, &\ldots, x_n, y_1, \ldots, y_{n-1}) \\ &= \sum \{ sign(\sigma) x_{\sigma(1)} y_1 x_{\sigma(2)} \cdots y_{n-1} x_{\sigma(n)} \mid \sigma \in S_n \}, \end{aligned}$$

which is multilinear and alternating in the variables $x_1, \ldots, x_n$.

Lemma 1.19. *Suppose that R is an A-algebra and $r_1, \ldots, r_{n+1} \in At_1 + \cdots + At_n$, an A-submodule of R generated by n elements. Then $s_{n+1}(r_1, \ldots, r_{n+1}) = 0$.*

Proof. Each r_i is an A-linear combination $\Sigma a_{ij} t_j$. The multilinearity of s_{n+1} implies that $s_{n+1}(r_1, \ldots, r_{n+1})$ is an A-linear combination of evaluations $s_{n+1}(t_{i_1}, \ldots, t_{i_{n+1}})$. Finally, the alternating property of s_{n+1} and the pigeon-hole principle imply that all such evaluations equal 0. $\square$

Exercise 1.20. Verify that if K has characteristic $p > 0$, then the Weyl algebra satisfies a PI. Hint: Show that x^p and y^p are central in the Weyl algebra, and that it is a finite module over $K[x^p, y^p]$. Then apply Lemma 1.19.

Definition 1.21. An A-algebra R has *bounded degree n* over A if for each $r \in R$ there is an A-submodule $M = At_1 + \cdots + At_n$ of R with n generators such that $A[r] \subseteq M$.

When A is a field, the definition simply means that every element of R is algebraic over A of degree $\leq n$.

Theorem 1.22. *Let R be an A-algebra.*

(a) *If $R \subseteq At_1 + \cdots + At_n$ for some $t_1, \ldots, t_n \in R$, then R satisfies $s_{n+1}(x_1, \ldots, x_{n+1})$.*

(b) *If R has bounded degree n, then R satisfies $s_{n+1}(x^n y, x^{n-1} y, \ldots, xy, y)$.*

Proof. (a) follows immediately from Lemma 1.19.
(b) Let $r, s \in R$. By hypothesis, there exist $t_1, \ldots, t_n \in R$ such that r^n, $r^{n-1}, \ldots, r, 1 \in At_1 + \cdots + At_n$. Then

$$r^n s, r^{n-1} s, \ldots, rs, s \in At_1 s + \cdots + At_n s.$$

By Lemma 1.19, $s_{n+1}(r^n s, r^{n-1} s, \ldots, rs, s) = 0$. $\square$

Exercise 1.23. Show that

$$s_{n+1}(x^n y, x^{n-1} y, \dots, xy, y) = s_n([x^n, y], [x^{n-1}, y], \dots, [x, y])y.$$

Chapter 2

The Amitsur–Levitzki Theorem

The main results of the next two chapters are the two most important theorems about the identities of $n \times n$ matrices, the Amitsur–Levitzki Theorem and the existence of central polynomials. Both theorems are important in both the structural and combinatorial sides of PI-theory, and will be proved here as well as in Drensky's course on combinatorial PI-theory. Since there are four essentially different proofs of the Amitsur-Levitzki Theorem and many constructions of central polynomials, Drensky and I will give different proofs of both results.

The original 1950 proof of S.A. Amitsur and J. Levitzki [AL] is a combinatorial inductive argument. A streamlined version can be found in [P, p. 175]. The 1958 proof of B. Kostant [Ko] is (in effect) based on the theory of *trace identities* and the theory of Frobenius relating the representations of the symmetric and alternating groups, although trace identities were not formally defined until 1974-76, by Y. P. Razmyslov [R2] and C. Procesi [Pr3]. The 1974 proof of Razmyslov is directly based on multilinearizing the Cayley–Hamilton Theorem, and is the most natural proof. The 1976 proof of S. Rosset [Ro] is the shortest and most ingenious, and uses the exterior (or Grassmann) algebra. All of the proofs except the combinatorial one depend on the Cayley–Hamilton Theorem.

We present Rosset's proof. We begin with a brief discussion of the exterior algebra, which is itself a PI-ring (see Example 5 in 1.16) and is important in combinatorial PI-theory, and then review some basic facts about commutative symmetric polynomials.

Definition 2.1. Let K be a field and let V be a vector space over K with basis $\{v_\sigma \mid \sigma \in S\}$. The *exterior (or Grassmann) algebra* of V over K is the free algebra $K\langle v_\sigma \mid \sigma \in S \rangle$ modulo the two-sided ideal generated by $\{v_\sigma v_\tau + v_\tau v_\sigma, \, v_\sigma^2 \mid \sigma, \tau \in S\}$. It is denoted $E(V)$.

The next lemma records some easily established properties of $E(V)$.

Lemma 2.2. *Let K be a field, and let V be a vector space over K with a finite basis $\{v_1, \ldots, v_n\}$ or a countable basis $\{v_1, v_2, \ldots\}$. Then*

(a) *The monomials $\{v_{i_1} \cdots v_{i_s} \mid i_1 < i_2 < \cdots < i_s\}$ (including the empty monomial 1) form a K-basis for $E(V)$ over K. Thus $E(V)$ has dimension 2^n over K if V has finite dimension n over K.*

(b) *The monomials of even degree in (a) are central in $E(V)$.*

(c) *The polynomial $f(x, y, z) = [[x, y], z] = (xy - yx)z - z(xy - yx)$ is a PI for $E(V)$.*

Proof. (a) and (b) are routine calculations. For (c), note that since $f(x, y, z)$ is multilinear, we only have to verify that it vanishes when we substitute monomials from (a) in f, by Lemma 1.11. If either u or v are monomials of even degree, then $[u, v] = 0$, since monomials of even degree are central. If both u and v have odd degree, then uv is zero or a monomial of even degree, and $[u, v] = uv - vu = 2uv$, which is central. In all cases $[[u, v], w] = 0$ for any $w \in E(V)$. $\square$

Definition 2.3. Let K be a field, and let $K[y_1, \ldots, y_n]$ be a commutative polynomial ring over K in independent indeterminates $y_1, \ldots, y_n$. A function $f(y_1, \ldots, y_n)$ is *symmetric* if $f(y_{\sigma(1)}, \ldots, y_{\sigma(n)}) = f(y_1, \ldots, y_n)$ for all $\sigma \in S_n$, where S_n is the symmetric group on n letters. For $i = 1, \ldots, n$, the *i-th elementary symmetric function* of $y_1, \ldots, y_n$ is $e_i(y_1, \ldots, y_n) = \sum \{y_{j_1} \cdots y_{j_i} \mid j_1 < j_2 < \cdots < j_i\}$, and the *i-th power symmetric function* of $y_1, \ldots, y_n$ is $p_i(y_1, \ldots, y_n) = y_1^i + \cdots + y_n^i$.

Theorem 2.4 (I. Newton). *Let $R = K[y_1, \ldots, y_n]$ be a commutative polynomial ring over a field K of characteristic zero, let S_n act on R by permuting variables, and let $R^{S_n} = \{r \in R \mid \sigma(r) = r \text{ for all } \sigma \in S_n\}$ be the fixed ring under this action. Then*

(a) $R^{S_n} = K[e_1, \ldots, e_n] = K[p_1, \ldots, p_n]$, *a polynomial ring in n independent variables.*

(b) $e_1, \ldots, e_n$ *and* $p_1, \ldots, p_n$ *generate the same ideal of R.* $\square$

By way of illustration, $p_1 = e_1$, $p_2 = e_1^2 - 2e_2$, $p_3 = e_1^3 - 3e_1e_2 + 3e_3$, $e_1 = p_1$, $2e_2 = p_1^2 - p_2$, $6e_3 = p_1^3 - 3p_1p_2 + 2p_3$, and these expressions are independent of the number of variables. The p_i lie in $\mathbb{Z}[e_i]$, but rational numbers are needed to express the e_i in terms of the p_i. The book of I.G. Macdonald [M] is a basic reference for symmetric functions.

Lemma 2.5. *Let K be a field of characteristic zero, let $Z \in M_n(K)$, and let $tr(U)$ denote the trace of a matrix U. If $tr(Z) = tr(Z^2) = \cdots = tr(Z^n) = 0$, then $Z^n = 0$.*

Proof. The characteristic polynomial of Z is

$$t^n - e_1 t^{n-1} + e_2 t^{n-2} - \cdots + (-1)^{n-1} e_{n-1} t + (-1)^n e_n,$$

where e_i is the i-th elementary symmetric function of the eigenvalues of Z. The Jordan canonical form (passing to an algebraic closure of K) shows that $tr(Z^i)$ is the i-th power symmetric function of the eigenvalues of Z. By Theorem 2.4(b), if the power symmetric functions $p_1, \ldots, p_n$ are all zero, so are the elementary symmetric functions $e_1, \ldots, e_n$, and then the characteristic polynomial of Z is just t^n. Since any $n \times n$ matrix satisfies its characteristic polynomial, $Z^n = 0$. $\square$

For a field K of characteristic $p > 0$, the $p \times p$ identity matrix shows that Lemma 2.5 is no longer valid.

Lemma 2.6. *Suppose that A is a commutative ring and $Z_1, \ldots, Z_{2r} \in M_n(A)$. Then $tr(s_{2r}(Z_1, \ldots, Z_{2r})) = 0$.*

Proof. The standard polynomial is a sum of monomials with coefficients ± 1. If we subdivide the monomials in $s_{2r}(Z_1, \ldots, Z_{2r})$ into equivalence classes of monomials which are cyclic permutations of each other, then all of the monomials in an equivalence class have the same trace, but half have a plus sign and half have a minus sign. Thus $tr(s_{2r}(Z_1, \ldots Z_{2r})) = 0$. $\square$

Theorem 2.7 (S.A. Amitsur–J. Levitzki [AL]). *Let A be a commutative ring. Then $s_{2n}(x_1, \ldots, x_{2n})$ is a polynomial identity for $M_n(A)$.*

Proof (S. Rosset [Ro]). It suffices to prove that s_{2n} is a PI for $M_n(K)$, where K is a field of characteristic zero. For then s_{2n} is a PI for $M_n(\mathbb{Z})$, and hence for $M_n(\mathbb{Z}[T])$, where T is any set of commuting variables over $\mathbb{Z}$, by Corollary 1.15. Then $M_n(A)$ also satisfies s_{2n}, since it is a homomorphic image of some $M_n(\mathbb{Z}[T])$.

Now let $U_1, \ldots, U_{2n} \in M_n(K)$, let V be a $2n$-dimensional vector space over K with basis $v_1, \ldots, v_{2n}$, let D be the subring of $E(V)$ generated by the monomials of even degree, and let

$$Y = U_1 v_1 + \cdots + U_{2n} v_{2n} \in M_n(E(V)),$$

$$Z = Y^2 = \sum \{(U_i U_j - U_j U_i) v_i v_j \mid 1 \le i, j \le 2n\}.$$

The first key observation is that $Z \in M_n(D)$, where D is a commutative ring. The second key observation is that for $k = 1, \ldots, n$,

$$Z^k = Y^{2k} = \sum \{s_{2k}(U_{i_1}, \ldots, U_{i_{2k}}) v_{i_1} \cdots v_{i_{2k}} \mid 1 \le i_1 < \cdots < i_{2k} \le 2n\},$$

and in particular

$$Z^n = Y^{2n} = s_{2n}(U_1, \ldots, U_{2n}) v_1 \cdots v_{2n}.$$

By Lemma 2.6, $tr(Z) = tr(Z^2) = \cdots = tr(Z^n) = 0$, so $Z^n = 0$, by Lemma 2.5, and $s_{2n}(U_1, \ldots, U_{2n}) v_1 \cdots v_{2n} = 0$. But then $s_{2n}(U_1, \ldots, U_{2n}) = 0$, since $M_n(E(V))$ is a free $M_n(K)$-module with basis $\{v_{i_1} \cdots v_{i_s} \mid i_1 < i_2 < \cdots < i_s\}$ (Lemma 2.2). $\square$

Theorem 2.8. *Let K be a field. Then $M_n(K)$ does not satisfy a polynomial identity of degree $\le 2n - 1$.*

Proof. If there were a PI of degree $\le 2n - 1$, then there would be a multilinear PI of degree $2n - 1$, by Theorem 1.14. Such a polynomial could be taken to have the form

$$f(x_1, \ldots, x_{2n-1})$$
$$= x_1 x_2 \cdots x_{2n-1} + \sum \{a(\sigma) x_{\sigma(1)} \cdots x_{\sigma(2n-1)} \mid \sigma \in S_{2n-1}, \sigma \ne 1, a(\sigma) \in K\}.$$

Since the "staircase" of $2n - 1$ matrix units $e_{11}, e_{12}, e_{22}, e_{23}, \ldots, e_{n-1,n-1}, e_{n-1,n}, e_{nn}$ has a nonzero product, e_{1n}, only when multiplied in the given order, it follows that $f(e_{11}, e_{12}, \ldots, e_{nn}) = e_{1n} \ne 0$. $\square$

Exercise 2.9. Show that if K is a field, then any multilinear polynomial $f(x_1, \ldots, x_{2n})$ satisfied by $M_n(K)$ is a scalar multiple of $s_{2n}(x_1, \ldots, x_{2n})$.

Chapter 3

Central Polynomials

We now turn to central polynomials for matrix rings. They were first constructed independently by E. Formanek [F1] in 1972 and Y.P. Razmyslov [R1] in 1973.

Definition 3.1. Let A be a commutative ring, and let R be an A-algebra with center C. A polynomial $f(x_1, \ldots, x_n) \in A\langle X \rangle$ is a *central polynomial* for R if

(a) For all $r_1, \ldots, r_n \in R$, $f(r_1, \ldots, r_n) \in C$.

(b) For some $r_1, \ldots, r_n \in R$, $f(r_1, \ldots, r_n) \neq 0$,

(c) The constant term of $f(x_1, \ldots, x_n)$ equals 0.

Part (a) of the definition can be replaced by

(a') $f(x_1, \ldots, x_n)x_{n+1} - x_{n+1}f(x_1, \ldots, x_n)$ is a polynomial identity for R.

Parts (b) and (c) of the definition are included so that polynomial identities and constant polynomials are not central polynomials.

The set of central polynomials for R (even with 0 included) do not form an ideal in $A\langle X \rangle$, but the additive abelian group generated by A and the set of central polynomials for R forms a subring of $A\langle X \rangle$. Moreover, any endomorphism of $A\langle X \rangle$ carries this subring into itself.

Examples, where K is a field 3.2.

(a) x is a central polynomial for commutative rings.

(b) For any vector space V over K, $[x, y] = xy - yx$ is a central polynomial for the exterior algebra $E(V)$.

(c) $[x, y]^2$ is a central polynomial for $M_2(K)$.

(d) $[x_1, y_1][x_2, y_2] \cdots [x_{n-1}, y_{n-1}]$ is a central polynomial for the ring of upper triangular $(2n + 1) \times (2n + 1)$ matrices over K with constant diagonal.

(e) The ring of upper triangular $n \times n$ matrices over K ($n \geq 2$, K infinite) has no central polynomials.

(f) If R does not satisfy a PI, then R has no central polynomials.

My favorite construction of central polynomials is my own, which is also the shortest. However, the polynomial has the disadvantage of not being multilinear.

Definition 3.3. An $n \times n$ *generic matrix* over a ring R is an $n \times n$ matrix whose entries are independent commuting indeterminates over R.

If $U = (u_{ij})$ is an $n \times n$ generic matrix over a field K, then the eigenvalues of U are distinct, so U is diagonalizable as a matrix over the algebraic closure of $K(u_{ij})$. This observation will be used in the proof below, and later in Chapter 12, when we introduce the *ring of generic matrices*, which is the K-algebra generated by a set of generic matrices with independent entries.

Theorem 3.4 (E. Formanek [F1]). *For each n, there exists a polynomial $F_n = F(x, y_1, \ldots, y_n) \in \mathbb{Z}\langle x, y_1, \ldots, y_n \rangle$ with the following properties.*

(a) *F is homogeneous of degree $n^2 - n$ in x and multilinear in $y_1, \ldots, y_n$.*

(b) *If K is a field, F is a central polynomial for $M_n(K)$.*

Proof. We first construct F and show that all evaluations of F on $M_n(K)$ are scalar matrices. For a given polynomial F, in order to show that all evaluations of F on $M_n(K)$ are scalar matrices, it suffices to show that any evaluation $F(X, Y_1, \ldots, Y_n)$ is a scalar matrix, when $X = (x_{ij})$ is an $n \times n$ generic matrix over K and $Y_1, \ldots, Y_n$ are arbitrary elements of $M_n(K)$.

If $\overline{K}$ is an algebraic closure of $K(x_{ij})$, there is a matrix $P \in M_n(\overline{K})$ such that PXP^{-1} is diagonal. Since

$$PF(X, Y_1, \ldots, Y_n)P^{-1} = F(PXP^{-1}, PY_1P^{-1}, \ldots, PY_nP^{-1}),$$

it suffices to show that $F(X, Y_1, \ldots, Y_n)$ is scalar when X is diagonal and $Y_1, \ldots, Y_n$ are arbitrary elements of $M_n(\overline{K})$. Then, using the multilinearity of f in the variables $y_1, \ldots y_n$ (as in the proof of Lemma 1.11) and noting that the matrix units $\{e_{ij}\}$ span $M_n(\overline{K})$ over $\overline{K}$, we have:

(∗)   If $F(X, Y_1, \ldots, Y_n)$ is a scalar matrix whenever X is a diagonal matrix and $Y_1 = e_{i_1 j_1}, \ldots, Y_n = e_{i_n j_n}$ are matrix units, then it is a scalar matrix for any $X, Y_1, \ldots, Y_n \in M_n(K)$.

We now describe $F(x, y_1, \ldots, y_n)$ explicitly. Let $K[u_1, \ldots, u_{n+1}]$ be a commutative polynomial ring over K in $n+1$ variables, and let φ be the K-linear map (not a ring homomorphism)

$$\varphi : K[u_1, \ldots, u_{n+1}] \to K\langle x, y_1, \ldots, y_n \rangle$$

defined on monomials by

$$\varphi(u_1^{a_1} \cdots u_{n+1}^{a_{n+1}}) = x^{a_1} y_1 x^{a_2} y_2 x^{a_3} y_3 \cdots y_{n-1} x^{a_n} y_n x^{a_{n+1}}.$$

Set $G(x, y_1, \ldots, y_n) = \varphi(g(u_1, \ldots, u_{n+1}))$, where

$$g(u_1, \ldots, u_{n+1}) = \prod_{2 \leq i \leq n} (u_1 - u_i)(u_{n+1} - u_i) \prod_{2 \leq j < k \leq n} (u_j - u_k)^2.$$

The desired central polynomial is

$$F(x, y_1, \ldots, y_n)$$
$$= G(x, y_1, \ldots, y_n) + G(x, y_2, \ldots, y_n, y_1) + \cdots + G(x, y_n, y_1, \ldots, y_{n-1}).$$

By (*) above, we need to show that $F(X, Y_1, \ldots, Y_n)$ is scalar when $X = v_1 e_{11} + \cdots + v_n e_{nn}$ is a diagonal matrix and $Y_1 = e_{i_1 j_1}, \ldots, Y_n = e_{i_n j_n}$ are matrix units. Note that $X e_{ij} = v_i e_{ij}$ and $e_{ij} X = v_j e_{ij}$, which implies that

$$X^{a_1} Y_1 X^{a_2} Y_2 \cdots X^{a_n} Y_n X^{a_{n+1}} = X^{a_1} e_{i_1 j_1} X^{a_2} e_{i_2 j_2} \cdots X^{a_n} e_{i_n j_n} X^{a_{n+1}}$$
$$= v_{i_1}^{a_1} v_{i_2}^{a_2} \cdots v_{i_n}^{a_n} v_{j_n}^{a_{n+1}} e_{i_1 j_1} \cdots e_{i_n j_n}.$$

Hence $G(X, e_{i_1 j_1}, \ldots, e_{i_n j_n}) = g(v_{i_1}, \ldots, v_{i_n}, v_{j_n}) e_{i_1 j_1} \cdots e_{i_n j_n}$. But

$$g(v_{i_1}, \ldots, v_{i_n}, v_{j_n}) = \prod_{2 \leq r \leq n} (v_{i_1} - v_{i_r})(v_{j_n} - v_{i_r}) \prod_{2 \leq s < t \leq n} (v_{i_s} - v_{i_t})^2.$$

Thus $g(v_{i_1} \ldots, v_{i_n}, v_{j_n}) = 0$ unless

(1) $(i_1, \ldots, i_n)$ is a permutation of $(1, \ldots, n)$ and $i_1 = j_n$.

Note that if (1) holds, then

(2) $g(v_{i_1}, \ldots, v_{i_n}, v_{j_n}) = \prod_{1 \leq s < t \leq n} (v_s - v_t)^2 = \Delta,$

where $\Delta = \Delta(v_1, \ldots v_n)$ is the discriminant of $v_1, \ldots, v_n$. Furthermore

(3) $e_{i_1 j_1} e_{i_2 j_2} \cdots e_{i_n j_n} = 0$ unless $j_1 = i_2, j_2 = i_3, \ldots, j_{n-1} = i_n$.

Let us call the sequence of matrix units $e_{i_1 j_1}, \ldots, e_{i_n j_n}$ a *cycle* if they have the form $e_{i_1 i_2}, e_{i_2 i_3}, e_{i_3 i_4}, \ldots, e_{i_{n-1} i_n}, e_{i_n i_1}$, where $i_1, \ldots, i_n$ is a permutation of $1, \ldots, n$. (1)–(3) show that

$$G(X, e_{i_1 j_1}, \ldots, e_{i_n j_n}) = \begin{cases} \Delta \cdot e_{i_1 i_1} & \text{if the matrix units are a cycle;} \\ 0 & \text{if the matrix units are not a cycle.} \end{cases}$$

Thus

$$F(X, e_{i_1 j_1}, \ldots, e_{i_n j_n}) = \begin{cases} \Delta \cdot I & \text{if the matrix units are a cycle;} \\ 0 & \text{if the matrix units are not a cycle.} \end{cases}$$

By (*), we have shown that for all $X, Y_1, \ldots, Y_n \in M_n(K)$, $F(X, Y_1, \ldots, Y_n)$ is a scalar matrix. It remains to show that F takes on nonzero values. This is clear from the construction if K has at least n elements. In general, it suffices to observe that there is always $X \in M_n(K)$ with distinct eigenvalues. Such an X is diagonalizable over the algebraic closure $\overline{K}$ of K, so there are $Y_1, \ldots, Y_n \in M_n(\overline{K})$ with $F(X, Y_1, \ldots, Y_n) \neq 0$. Finally, the multilinearity of F in $y_1, \ldots, y_n$ implies that $Y_1, \ldots, Y_n$ can be found in $M_n(K)$. $\square$

Note that evaluations of the intermediate polynomial G commute with X. In other words, $[x, G(x, y_1, \ldots, y_n)]$ is a PI for $M_n(K)$, but for $n \geq 2$, $G(x, y_1, \ldots, y_n)$ is not a central polynomial for $M_n(K)$.

The polynomial F can be multilinearized by taking the multilinear component of $F(x_1 + \cdots + x_{n^2-n}, y_1, \ldots, y_n)$. Evaluations of the resulting polynomial on $M_n(K)$ are always central, but it is not known if there are nonzero evaluations when K has characteristic $p > 0$.

The quite easy Theorem 2.8 showed that $2n$, the degree of s_{2n}, is the minimal degree of a PI satisfied by $M_n(K)$. In contrast, the minimal degree for a central polynomial for $M_n(K)$ is an open problem which seems to be difficult. The only polynomial function consistent with known data is $1/2(n^2 + 3n - 2)$. This gives the minimal degree for $n \leq 3$ [DK], and it is also known that there is a central polynomial of degree 13 for $M_4(K)$ [DP], but that is the extent of current knowledge.

The following observation is due to C. Procesi.

Lemma 3.5. *Let $f \in \mathbb{Z}\langle X \rangle$ be a central polynomial for $M_n(\mathbb{Z})$. Then f is a polynomial identity for $M_r(\mathbb{Z})$ for $r < n$.*

Proof. Consider the (not unitary) embedding of $M_r(\mathbb{Z})$ into the upper left corner of $M_n(\mathbb{Z})$. $\qquad\square$

Chapter 4

Kaplansky's Theorem

The first paper in which the notion of polynomial identity appears that I know of is a 1922 article of M. Dehn [D], where his goal was to generalize Pappus' Theorem from projective geometry. (His program was carried out by S.A. Amitsur in 1966 [A4] using *rational identities*.) During the next twenty-five years, there were only a couple of significant results in the area. In 1937, W. Wagner [W] constructed polynomial identities satisfied by $n \times n$ matrices, and in 1943, M. Hall [H] proved the following theorem.

Theorem 4.1 (M. Hall [H]). *Let D be a division algebra in which $[x, y]^2$ is central for all $x, y \in D$. Then either D is commutative, or D is four-dimensional over its center.* □

Hall's theorem was about rings, but one can see from the titles that the topic of his paper, as well as those of Dehn and Wagner, was projective geometry. The term *polynomial identity* first appeared in the 1948 paper of I. Kaplansky [K], and *rings with polynomial identity* as a field of study owes its existence to this paper. His main result, which generalized Hall's theorem, is the foundation for the structure theory of PI-rings.

As often happens in mathematics, one good theorem leads to others, and Kaplansky's proof depends on the Chevalley–Jacobson Density Theorem, which at that time was only a few years old. We begin by reviewing this theorem and one of its consequences.

Definition 4.2. A ring R is (left) *primitive* if there is a simple faithful left R-module.

Examples, where K is a field 4.3.

1. There exist left primitive rings which are not right primitive.

2. A commutative ring is primitive if and only if it is a field.

3. A division ring is primitive.

4. The free algebra $K\langle X \rangle$ is primitive, but the polynomial ring $K[X]$ is not primitive, since it is commutative, but not a field.

5. Let V be a vector space over K, and let E be the two-sided ideal of $End_K(V)$ consisting of those $\alpha \in End_K(V)$ such that $\alpha(V)$ is finite-dimensional over K. Then $K \cdot I + E$ is primitive, where $I : V \to V$ is the identity map. (If we allowed rings without unit, E would already be primitive.) When V is finite-dimensional over K, $K \cdot I + E = End_K(V)$, but $K \cdot I + E$ is strictly smaller than $End_K(V)$ when V is infinite-dimensional over K.

6. A ring R is primitive if and only if $M_n(R)$ is primitive.

Definition 4.4. If R and D are rings, an abelian group M is an *R-D-bimodule* if

(a) M is a left R-module and a right D-module.

(b) If $r \in R$, $d \in D$, and $m \in M$, then $(rm)d = r(md)$.

(We write $_RM$, M_D, and $_RM_D$ when we wish to emphasize that M is, respectively, a left R-module, a right D-module, or an R-D-bimodule.)

If $_RM$ is a simple left R-module, then $D = End_R(_RM)$ is a division ring (Schur's Lemma), and M is an R-D-bimodule, provided that we regard elements of D as endomorphisms acting on M from the right.

Definition 4.5. Let R be a ring, let D be a division ring, and let $M = _RM_D$ be an R-D-bimodule which is faithful as an R-module. Suppose that whenever $m_1, \ldots, m_k$ is a finite set of D-linearly independent elements of M, and $n_1, \ldots, n_k$ are any elements of M, then there is an $r \in R$ such that $rm_i = n_i$ for $i = 1, \ldots, k$. Then R is said to be a *dense ring of linear transformations* on M over D (or R is a *dense subring of* $End_D(M_D)$).

The definition of R-D-bimodule says that there is a ring homomorphism $R \to End_D(M_D)$. This homomorphism is one-to-one if and only if M is faithful as a left R-module. If M is finite-dimensional as a right D-module, then the only dense subring of $End_D(M_D)$ is $End_D(M_D)$ itself, and $End_D(M_D)$ is isomorphic to $M_n(D)$, where $n = dim_D(M_D)$.

Suppose that R is a dense subring of $End_D(M_D)$. If $m, m_1 \in M$ and $m \neq 0$, then there is an $r \in R$ such that $rm = m_1$. Hence $Rm = M$ for any nonzero $m \in M$, so M is a simple R-module. By the definition of dense, M is a faithful R-module. Hence R is primitive, since it has a simple faithful R-module. The converse of this observation is the Chevalley–Jacobson Density Theorem.

Theorem 4.6 (Chevalley–Jacobson Density Theorem [J, p. 199]). *Suppose that R is a primitive ring and M is a simple faithful R-module, and let $D = End_R(_RM)$. Then R is a dense subring of $End_D(M_D)$.* $\square$

An essential fact used in the proof of Kaplansky's Theorem is that if a division algebra D with center K is infinite-dimensional over K, then it does not satisfy a PI. The proof depends on the density theorem as well as the theory of central simple algebras, which can be found in Chapter 4 of [He1].

Lemma 4.7. *Let R be a primitive ring, and let $D = End_R(M)$, where M is a simple faithful R-module. Then either*

(a) *For some integer n, $dim_D(M_D) = n$, and R is isomorphic to $M_n(D)$; or*

(b) *For every positive integer n, there exists a subring R_n of R which maps homomorphically onto $M_n(D)$.*

Proof. If $dim_D(M_D) = n < \infty$, then (a) holds by the density theorem. If $dim_D(M_D) = \infty$ and n is any positive integer, let V_n be an n-dimensional D-subspace of M_D, and let $R_n = \{r \in R \mid rV_n \subseteq V_n\}$. There is an obvious ring homomorphism $R_n \to End_D(V_n) \cong M_n(D)$, which is onto by the density theorem. $\qquad\square$

Definition 4.8. A ring R is *simple* if 0 and R are the only two-sided ideals of R. If R is a simple ring whose center is a field K, then R is called a *central simple K-algebra*.

Exercise 4.9. Show that the center of a simple ring is a field.

Lemma 4.10 ([He1, Lemma 4.1.1, p. 90]). *If R is a central simple K-algebra and L is an extension field of K, then $R \otimes_K L$ is a central simple L-algebra.* $\qquad\square$

Definition 4.11. A subfield L of a ring R is called a *maximal subfield* of R if it satisfies: If $L \subseteq L' \subseteq R$ and L' is a subfield of R, then $L = L'$.

If R contains a field, then it contains a maximal subfield, by Zorn's Lemma.

Lemma 4.12 ([He1, Theorem 4.2.1, p. 95]). *Let D be a division algebra with center K, and let L be a maximal subfield of D. Then $D \otimes_K L$ is a dense ring of linear transformations of $End_L(D_L)$, and either*

(a) *For some positive integer n, $dim_K(L) = n$, $dim_K(D) = n^2$, and $End_L(D_L) \cong M_n(L)$; or*

(b) *$dim_K(L) = \infty$ and $dim_L(D) = \infty$.* $\qquad\square$

Theorem 4.13 (I. Kaplansky [K]). *Let R be a primitive ring which satisfies a polynomial identity. Then R is isomorphic to $M_n(D)$, where D is a division algebra finite-dimensional over its center K. Equivalently, R is a central simple algebra which is finite-dimensional over its center K.*

Proof. By Theorem 1.14, R satisfies a proper multilinear polynomial f.

Let M be a simple faithful R-module and let $D = End_R(_RM)$. By Theorem 2.8, there is no polynomial which is a PI for $M_n(D)$ for all n. Hence possibility (b) in Lemma 4.7 is excluded. Thus (a) holds, and R is isomorphic to $M_n(D)$ for some n.

Let K be the center of D, and let L be a maximal subfield of D. Since D contains an L-basis for $D \otimes_K L$ and f is multilinear, f is a PI for $D \otimes_K L$, by Lemma 1.11.

By Lemma 4.12, $D \otimes_K L$ is a dense ring of linear transformations of $End_L(D_L)$. The argument of the preceding paragraph excludes possibility (b) of Lemma 4.12. Thus Lemma 4.12(a) holds, and $dim_K(D) = n^2$ for some positive integer n. $\qquad\square$

Kaplansky's Theorem motivates a definition, based on the following lemma.

Lemma 4.14. *Let R be a central simple algebra finite-dimensional over its center K. Then there is a positive integer n such that*

(a) *$dim_K(R) = n^2$.*

(b) *Every polynomial identity satisfied by $M_n(\mathbb{Z})$ is satisfied by R. In particular, the standard polynomial s_{2n} is a PI for R.*

(c) *No polynomial of degree $\leq 2n - 1$ is a PI for R.*

(d) *The Formanek polynomial $F_n = F(x, y_1, \ldots, y_n)$ of Theorem 3.4 is a central polynomial for R.*

Proof. By Lemma 4.7, $R \cong M_r(D)$ for some integer r, where D is a division algebra finite-dimensional over K. By Lemma 4.12, $D \otimes_K L \cong M_s(L)$, where L is a maximal subfield of D and $dim_K(L) = s$. Then

$$R \otimes_K L \cong M_r(D) \otimes_K L \cong M_r(D \otimes_K L) \cong M_r(M_s(L)) \cong M_{rs}(L),$$

and so $dim_K(R) = dim_L(R \otimes_K L) = r^2 s^2$. This proves (a), with $n = rs$.

By Lemma 1.11, R and $R \otimes_K L$ satisfy the same multilinear polynomials. Since $M_n(L)$ satisfies all the polynomial identities satisfied by $M_n(\mathbb{Z})$ and does not satisfy a polynomial of degree $\leq 2n - 1$ by Theorem 2.8, (b) and (c) hold.

By Theorem 3.4, F_n is a central polynomial for $M_n(E)$, where E is any field. If K is finite, then Wedderburn's Theorem (a finite division ring is a field) implies that $R = M_n(K)$. If K is infinite, then Theorem 1.9 says that R and $R \otimes_K L \cong M_n(L)$ satisfy the same polynomials in $K\langle X \rangle$. In both cases F_n is a central polynomial for R, which proves (d). $\square$

Definition 4.15. A ring R has *PI-degree n* if it satisfies (b), (c) and (d) of Lemma 4.14.

Thus a central simple algebra of dimension n^2 over its center has PI-degree n. However, other rings have PI-degree n. For example, if A is a commutative ring, then $M_n(A)$ has PI-degree n.

Chapter 5

Theorems of Amitsur and Levitzki on Radicals

In the 1940s and 1950s the main topics in ring theory were radical theory, nilpotence, algebraicity and local finiteness. It is no surprise that the early work of S.A. Amitsur, J. Levitzki, and I. Kaplansky on PI-rings addressed these topics. They were able to give positive answers for PI-rings to questions which remained open or had negative answers for rings in general.

For the most part, these lectures will treat developments which go in other directions. However, in order to prove Posner's theorem, we need two important results of Amitsur from this period. The first concerns nil PI-rings, the second is a general result about the Jacobson radical of polynomial extensions. We also prove that every PI-algebra satisfies a power of the standard identity, another result of Amitsur.

In this chapter we will need to work with rings which may not have a unit. To be precise, polynomial identities should now be defined in terms of a free associative algebra without a unit, but we will not formally do this. A definition and discussion can be found in Chapter 1, Section 1.2 of Drensky's part of the book.

Definition 5.1. A *rng* is a ring which may not have a unit element. A rng R is a *domain* if $ab \neq 0$ whenever a and b are nonzero elements of R. A rng R is *prime* if $IJ \neq 0$ whenever I and J are nonzero two-sided ideals of R.

Examples 5.2. 1. A commutative rng is prime if and only if it is a domain.

2. If R is a prime rng, then $M_n(R)$ is a prime rng. But $M_n(R)$ is not a domain unless R is a domain and $n = 1$.

Exercise 5.3. Show that if a prime rng R is not a domain, then there is a nonzero $r \in R$ such that $r^2 = 0$.

Lemma 5.4. *Suppose that R is a prime rng and that L is a left ideal of R. Let*

$$r(L) = \{r \in R \mid Lr = 0\}.$$

Then

(a) $M = r(L) \cap L$ *is a two-sided ideal in the rng L.*

(b) L/M *is a prime rng.*

Proof. (a) $L \cdot M = 0$, so $L \cdot M \subseteq M$. Furthermore, $L \cdot (M \cdot L) = 0$, so $M \cdot L \subseteq r(L) \cap L = M$.

  (b) Suppose that I and J are two-sided ideals of L and $IJ \subseteq M$. Then

$$(LI + LIR)(LJ + LJR) \subseteq LIJ + LIJR \subseteq LM + LMR = 0,$$

so $LI + LIR = 0$ or $LJ + LJR = 0$, since they are two-sided ideals in R, a prime rng. Thus $I \subseteq M$ or $J \subseteq M$.   □

In proving the next theorem, we use the fact that Theorem 1.14 is valid for rngs – i.e., every PI-rng satisfies a proper multilinear polynomial, provided of course, that when identities lie in $A\langle X \rangle$, the underlying commutative ring A is assumed to have a unit. (If A had no unit, then there would be no proper polynomials in $A\langle X \rangle$.) Note that for a rng R to be an algebra over a commutative ring A means only that R is an A-module satisfying the compatibility condition: $a(rs) = r(as) = (ar)s$ for all $a \in A$, $r, s \in R$. It does not mean that there is a ring homomorphism from A to the center of R.

Definition 5.5. A subset S of a rng R is *nil* if every element of S is nilpotent.

Theorem 5.6 (J. Levitzki [L]). *If R is a prime PI-rng, then R contains no nonzero nil ideals.*

Proof. Let R satisfy the proper multilinear polynomial

$$f(x_1, \ldots, x_n) = x_1 x_2 \cdots x_n + \sum \{a(\sigma) x_{\sigma(1)} x_{\sigma(2)} \cdots x_{\sigma(n)} \mid \sigma \in S_n, \ \sigma \neq 1\}$$

in $\mathbb{Z}\langle X \rangle$ (or $A\langle X \rangle$, if R is an A-algebra).

  The proof is by induction on n, the degree of f. The case $n = 1$ is clear, since then $f(x_1) = x_1$ and $R = 0$.

  For the inductive step, suppose that the result is true whenever a prime PI-rng satisfies a proper multilinear PI of degree $n - 1$. We will assume that R contains a nonzero nil ideal N, and show that this leads to a contradiction.

  Choose a nonzero $a \in N$ such that $a^2 = 0$, and let $L = Ra$, a left ideal of R. By Lemma 5.4(b), $R_1 = L/(r(L) \cap L)$ is a prime rng. We next show that R_1 satisfies a proper multilinear PI of degree $n - 1$. Let

$$f(x_1, \ldots, x_n) = x_1 g(x_2, \ldots, x_n) + h(x_1, \ldots, x_n),$$

where no monomial in $h(x_1, \ldots, x_n)$ begins with x_1. For any $r_1, \ldots, r_n \in R$,

$$\begin{aligned} 0 = f(ar_1, r_2 a, \ldots, r_n a) &= ar_1 g(r_2 a, \ldots, r_n a) + h(ar_1, r_2 a, \ldots r_n a) \\ &= ar_1 g(r_2 a, \ldots, r_n a). \end{aligned}$$

Hence $aRg(r_2 a, \ldots, r_n a) = 0$, and replacing ar_1 by a shows in the same way that $ag(r_2 a, \ldots, r_n a) = 0$. This implies that the product of the two-sided ideals of R generated by, respectively, a and $g(r_2 a, \ldots, r_n a)$ equals 0, so $g(r_2 a, \ldots, r_n a) = 0$, since R is a prime rng. Thus $g(x_2, \ldots, x_n)$ is a proper multilinear PI for L, and hence also for R_1, as claimed.

By the inductive hypothesis, R_1 has no nonzero nil ideals. Since $L = Ra \subseteq N$, L is nil, and so R_1 is a nil ideal of itself. Thus $0 = R_1 = L/(r(L) \cap L)$, so $RaRa = L^2 = 0$, and so $Ra = 0$, since R is a prime rng. But then the square of the two-sided ideal of R generated by a is zero, which contradicts the hypothesis that R is a prime rng. This contradiction completes the inductive step, and the proof. $\qquad\square$

Definition 5.7. A ring R is the *subdirect product* of the rings $\{R_\alpha \mid \alpha \in \mathcal{A}\}$ if

(a) Each R_α is isomorphic to a homomorphic image R/M_α of R, where each M_α is a two-sided ideal of R.

(b) The diagonal map $R \to \prod\{R_\alpha \mid \alpha \in \mathcal{A}\}$ is one-to-one.

Definition 5.8. A ring R is *semiprimitive* if it is a subdirect product of primitive rings. It is *semiprime* if it is a subdirect product of prime rings.

Hard Exercise 5.9. Show that a ring R is semiprime if and only if the square of a nonzero two-sided ideal of R is nonzero.

Definition 5.10. The *Jacobson Radical*, $\mathcal{J}(R)$ of a ring R is the intersection of the maximal left ideals of R.

The next theorem gives a few of the many characterizations of the Jacobson Radical.

Theorem 5.11. *Let R be a ring.*

(a) $\mathcal{J}(R)$ *is the intersection of the two-sided ideals M of R such that R/M is left primitive.*

(b) $\mathcal{J}(R) = \{r \in R \mid 1 - ar \text{ is invertible for all } a \in R \}$.

(c) $\mathcal{J}(R) = \{r \in R \mid 1 - ra \text{ is invertible for all } a \in R \}$. $\qquad\square$

Definition 5.12. Let R be a ring. The *ring of formal power series over R*, is the set of expressions $r_0 + r_1 t + r_2 t^2 + \ldots$ with ring operations compatible with those of the polynomial ring $R[t]$. It is denoted $R[[t]]$.

Lemma 5.13. *Let R be a ring, and let $r = r_0 + r_1 t + \cdots + r_n t^n \in R[t]$, where $r_0, r_1, \ldots, r_n$ commute. Then r is invertible in $R[t]$ if and only if r_0 is invertible in R and $r_1, \ldots, r_n$ are nilpotent.*

Proof. Suppose that r is invertible in $R[t]$. Clearly, r_0 is invertible in R. Letting R_0 be the commutative subring of R generated by $r_0^{-1}, r_0, r_1, \ldots, r_n$, we see that r is invertible in $R_0[[t]]$. By the uniqueness of inverses in $R[[t]]$, r^{-1} is an element of $R[t] \cap R_0[[t]] = R_0[t]$. But it is a standard result from commutative algebra [AM, Exercise 2(1), p. 10] that the lemma holds for commutative rings. This implies that $r_1, \ldots, r_n$ are nilpotent. The converse is trivial. $\qquad\square$

Note that a ring R is semiprimitive if and only if $\mathcal{J}(R) = 0$.

Theorem 5.14 (S.A. Amitsur [A3]). *Let R be a ring which contains no nil ideals, and let t be a central indeterminate over R. Then $R[t]$ is semiprimitive.*

Proof. We will show that if $\mathcal{J}(R[t]) \neq 0$, then R contains a nonzero nil ideal. Assuming that $\mathcal{J}(R[t]) \neq 0$, let n be the minimal degree of a nonzero element of $\mathcal{J}(R[t])$, and set

$$I = \{r_n \in R \mid \text{ some } r = r_0 + r_1 t + \cdots + r_n t^n \text{ lies in } \mathcal{J}(R[t])\}.$$

Then I is a nonzero two-sided ideal of R. The minimal degree, n, of $r \in \mathcal{J}(R[t])$ shows that $[r_n, r] = 0$, so $[r_n, r_i] = 0$ for $i = 0, 1, \ldots, n$. Continuing this argument inductively with $r_{n-1}, r_{n-2}, \ldots$ shows that $r_0, r_1, \ldots, r_n$ all commute.

By Theorem 5.11, $1 - rt$ is invertible in $R[t]$, and then Lemma 5.13 implies that r_n is nilpotent. Thus I is a nonzero nil ideal of R. $\square$

Chapter 6

Posner's Theorem

It is elementary that a primitive commutative ring is a field. Kaplansky's Theorem replaces "commutative" with "satisfies a polynomial identity" and says that a primitive PI-ring is a finite-dimensional central simple algebra over a field.

The next basic structure theorem for PI-rings is Posner's Theorem, a localization theorem which likewise generalizes an elementary result about commutative rings:

A commutative domain has a quotient field.

A commutative ring is a domain if and only if it is a prime ring. In general, a prime ring is not a domain, although its center is. It turns out that for PI-rings, the right generalization of the above result has the hypothesis "prime ring" rather than "domain".

Posner's Theorem says that a prime PI-ring R has a quotient ring $Q(R)$ which is a primitive PI-ring satisfying the same polynomial identities as R. It was originally proved by E.C. Posner [Po] using Goldie's Theorem and noncommutative localization via the Ore condition. The modern treatment based on central polynomials shows that $Q(R)$ can be obtained by central localization. The approach we follow is due to L.H. Rowen [Rw1]. It uses Theorems 5.6 and 5.14 of Levitzki and Amitsur to give an efficient and well-motivated proof of Posner's Theorem.

Lemma 6.1. *If R is a semiprime PI-ring and t is a central indeterminate, then $R[t]$ is a semiprimitive PI-ring.*

Proof. By Corollary 1.15, $R[t]$ is a PI-ring. Let $R \to \prod R_\alpha$ express R as a subdirect product of prime PI-rings. Then $R[t] \to \prod R_\alpha[t]$ expresses $R[t]$ as a subdirect product of semiprimitive rings, since each $R_\alpha[t]$ is semiprimitive, by Theorems 5.6 and 5.14. Thus $R[t]$ is semprimitive. □

Theorem 6.2 (L.H. Rowen [Rw1]). *Let R be a semiprime PI-ring with center C. If J is a nonzero two-sided ideal of R, then $J \cap C \neq 0$.*

Proof. Let t be a central indeterminate over R. By Lemma 6.1, $R[t]$ is a semiprimitive PI-ring. It is easy to see that $C[t]$ is the center of $R[t]$, and that for any two-sided ideal J of R, $(J[t] \cap C[t]) = (J \cap C)[t]$. Hence we may assume that R is semiprimitive.

Assuming now that R is semiprimitive and satisfies a PI of degree d, let $R \to \prod \{R_\alpha \mid \alpha \in \mathcal{A}\}$ express R as a subdirect product of primitive PI-rings. Each R_α is central simple of some dimension n_α^2 over its center, where n_α is the

PI-degree of R_α. Hence $2n_\alpha \leq d$, since R_α satisfies no PI of degree $< 2n_\alpha$, by Lemma 4.14(c). Thus the integers $\{n_\alpha \mid \alpha \in \mathcal{A}\}$ are bounded above.

Consider the image of J under the diagonal map $R \to \prod\{R_\alpha \mid \alpha \in \mathcal{A}\}$. Since the R_α are simple rings, the projection of J on R_α is either 0 or R_α. Moreover, it is R_α for some α since $J \neq 0$. Let n be the maximum of the n_α for which the projection of J on R_α is R_α.

Let $F_n = F(x, y_1, \ldots, y_n)$ be the Formanek polynomial of Theorem 3.4. By Lemma 4.14(d), F_n is a central polynomial for all central simple algebras of PI-degree n. Consider the set of evaluations of f on J, projected onto R_α. When $n_\alpha > n$, the projection is zero, since the projection of J on R_α is zero. When $n_\alpha = n$, the projection is central, since f is a central polynomial for R_α. When $n_\alpha < n$, the projection is zero, since f is a PI for R_α (Lemma 3.5). Thus the set of evaluations of f on J is central in R. Moreover, there are some evaluations which are nonzero, since there are some R_α such that $n_\alpha = n$ and J projects onto R_α. These nonzero evaluations are nonzero elements of $J \cap C$. $\qquad\square$

Corollary 6.3 (L.H. Rowen [Rw1]). *Let R be a semiprime PI-ring with center C. If C is a field, then R is a finite-dimensional central simple C-algebra.*

Proof. If C is a field, then R is simple, by Theorem 6.2, and hence primitive. Then Kaplansky's Theorem (4.13) gives the conclusion. $\qquad\square$

Exercise 6.4. Let R be a prime ring with center C, and let $S = C - \{0\}$. Show that C is a domain and that its quotient field $S^{-1}C$ is the center of $S^{-1}R$.

Theorem 6.5 (E.C. Posner [Po]). *Let R be a prime PI-ring with center C, and let $S = C - \{0\}$. Then $S^{-1}R$ is a finite-dimensional central simple $S^{-1}C$-algebra. Moreover, R and $S^{-1}R$ satisfy exactly the same polynomial identities in $\mathbb{Z}\langle X \rangle$ (or $K\langle X \rangle$, if K is a field and R is a K-algebra).*

Proof. By exercise 6.4, $S^{-1}R$ is a prime ring whose center is the field $S^{-1}C$.

The proof that R and $S^{-1}R$ satisfy exactly the same polynomial identities in $\mathbb{Z}\langle X \rangle$ (or $K\langle X \rangle$) breaks into two cases. If C is finite, then it is already a field, and $R = S^{-1}R$. If C is infinite, then R and $R[T]$ satisfy the same polynomial identities, for any set T of central indeterminates, by Corollary 1.15. Since $S^{-1}R$ is a homomorphic image of $R[T]$ for a large enough set T, $S^{-1}R$ satisfies every PI satisfied by R. Conversely, since $R \subseteq S^{-1}R$, R satisfies every PI satisfied by $S^{-1}R$.

Since $S^{-1}R$ is a prime PI-ring whose center is a field, Corollary 6.3 implies that it is a finite-dimensional $S^{-1}C$-algebra. $\qquad\square$

One consequence of Posner's Theorem is that a prime PI-ring R has PI-degree n, where n^2 is the dimension of $S^{-1}R$ over its center.

Chapter 7

Every PI-ring Satisfies a Power of the Standard Identity

Using the fact that a prime PI-ring R satisfies s_{2n} if it has PI-degree n, we can prove a basic result of Amitsur which implies that every PI-algebra is a PI-ring - i.e., every PI-algebra satisfies a proper PI with integer coefficients.

Definition 7.1. The *prime radical* of a ring R is the intersection of the prime ideals of a ring R.

Lemma 7.2. *The prime radical of a ring R is a nil ideal.*

Proof. If x is not nilpotent, let $S = \{1, x, x^2, \dots \}$. By Zorn's Lemma the set

$$\{I \text{ an ideal of } R \mid I \cap S = \phi\}$$

contains maximal elements, and such maximal elements are prime ideals which do not contain x. Thus x is not in the prime radical of R. $\square$

Theorem 7.3 (S.A. Amitsur [A1]). *Let R be a PI-algebra over a ring A. Then R satisfies $s_{2n}(x_1, \dots, x_{2n})^m$ for some integers n and m.*

Proof. Let R satisfy a proper PI f of degree d. If R_0 is any A-algebra satisfying f, then any prime homomorphic image R_0/P of R_0 has PI-degree $n(R_0/P)$, where $2n(R_0/P) \le d$, by Lemma 4.14(c). Hence any such prime homomorphic image of R_0 satisfies $s_{2n}(x_1, \dots, x_{2n})$, where $2n$ is the least even integer $\ge d$. This implies that for any $r_1, \dots, r_{2n} \in R_0$, $s_{2n}(r_1, \dots, r_{2n})$ lies in the prime radical of R_0, so it is nilpotent, by Lemma 7.2.

It remains to show that there is an index of nilpotence which is independent of the particular elements $r_1, \dots, r_{2n} \in R$. There are two formally different but philosophically similar ways to do this.

Amitsur's argument: Let $S = \prod \{R_\alpha \mid \alpha \in \mathcal{A}\}$ be the direct product of copies R_α of R, indexed by the set $\mathcal{A}$ of all 2n-tuples $\alpha = (r_1, \dots, r_{2n})$, $r_i \in R$. Let $r_i(\alpha)$ denote the i-th coordinate of α, and let $u_1, \dots, u_{2n} \in R_0$ be defined by specifying that the α-coordinate of u_i is $r_i(\alpha)$. Since S satisfies the PI f, there is an integer m such that $s_{2n}(u_1, \dots, u_{2n})^m = 0$. Since the α-coordinate of $s_{2n}(u_1, \dots, u_{2n})^m$ is $s_{2n}(r_1(\alpha), \dots, r_{2n}(\alpha))^m$, this means that $s_{2n}(r_1, \dots, s_{2n})^m = 0$ for any $r_1, \dots, r_{2n} \in R$.

T-ideal argument: Let T be the T-ideal of $A\langle X \rangle$ generated by f, and let $S = A\langle X \rangle / T$. By the argument of the first paragraph of the proof, there are integers n and m such that $s_{2n}(\overline{x_1}, \ldots, \overline{x_{2n}})^m = 0$, where $\overline{x_i}$ denotes the image of $x_i \in X$ under the canonical homomorphism $A\langle X \rangle \to S$. For any $r_1, \ldots, r_{2n} \in R$, $s_{2n}(r_1, \ldots, r_{2n})^m$ is the image of $s_{2n}(\overline{x_1}, \ldots, \overline{x_{2n}})^m = 0$ under the homomorphism $S \to R$ induced by the specialization $\overline{x_i} \to r_i$. (I.e., the specialization $A\langle X \rangle \to R$ induced by $x_i \to r_i$ factors through S.) Thus $s_{2n}(r_1, \ldots, r_{2n})^m = 0$. □

Theorem 7.3 does not give any information on the size of n and m. By combinatorial means A. Regev [Re, Theorem 4.3] showed that if A is a field of characteristic zero and R satisfies a polynomial of degree d, then R satisfies $(\mathcal{S}_{d^2})^{d^8/4}$.

Chapter 8

Azumaya Algebras

We begin by reviewing, without proofs, some facts about projective modules.

Theorem 8.1. *Let R be a ring, and let M be an R-module. The following are equivalent.*

(a) *Every short exact sequence*

$$0 \to S \longrightarrow T \longrightarrow M \to 0$$

of R-modules splits.

(b) *M is a direct summand of a free R-module.* □

Definition 8.2. An R-module M is *projective* if it satisfies the equivalent conditions of Theorem 8.1.

Theorem 8.3. *If R is a local ring, then every projective module over R is free. (A ring R is local if $R/J(R)$ is a division ring, where $J(R)$ is the Jacobson radical of R.)* □

Notation 8.4. If A is a commutative ring and P is a prime ideal in A, then A_P denotes the localization of A at the multiplicative set $A - P$, where $A - P = \{a \in A \mid a \notin P\}$. If M is an A-module, then M_P denotes the A_P-module $A_P \otimes_A M$.

If A is a commutative ring, P is a prime ideal of A, and M is a projective A-module, then M_P is a projective module over the local ring A_P, and hence free. If M is finitely generated as an A-module, then M_P is finitely generated as an A_P-module, so it is free of finite rank. The rank is well-defined since if a free module of rank r is isomorphic to a free module of rank s, then $r = s$. (Warning: This is true since A is commutative. There exist noncommutative rings for which a free module of rank r is isomorphic to a free module of rank s, with $r \neq s$.)

Exercise 8.5. Let A be a commutative ring. Show that if r and s are positive integers and A^r and A^s are isomorphic as A-modules, then $r = s$. Hint: Note that $End_A(A^r)$ is isomorphic to $M_r(A)$ and use Lemma 4.14(b,c).

Note that the rank of M_P as an A_P-module is the same as the dimension of M_P/PM_P as a vector space over A_P/PA_P. If P is a maximal ideal, then $A_P/PA_P \cong A/P$. In general, A_P/PA_P is isomorphic to the field of quotients of A/P.

Definition 8.6. Let M be a finitely generated projective module over a commutative ring A. The *rank* of M is the function $rank_M : \{$ prime ideals of $A\} \to \mathbb{N}$ defined by $rank_M(P) = $ rank of M_P as an A_P-module.

If $rank_M(P) = r$ for all prime ideals P in A, then M is said to be a projective module of *constant rank* r.

Theorem 8.7. *Let M be a finitely generated projective module over a commutative ring A. Then there exist orthogonal idempotents $e_1, \ldots, e_k$ in A such that $e_1 + \cdots + e_k = 1$ and $e_r M$, regarded as an $e_r A$-module, has constant rank r.* □

An equivalent but fancier reformulation of Theorem 8.7 is that $rank_M$ is a continuous function when the set of prime ideals of A is given the *Zariski topology* (closed sets are the sets $\mathcal{V}(I) = \{$ prime ideals P of $A \mid P \supseteq I\}$, where I is a subset of A) and $\mathbb{N}$ is given the discrete topology.

We now introduce Azumaya algebras and develop some of their properties. Standard references for Azumaya algebras are [DI] and [OS]. Our convention that rings have unit elements has not been essential so far, but it is now, because Azumaya algebras are projective modules over their centers, and projective modules are only useful over rings with a unit. Azumaya algebras are a generalization of central simple algebras in which the center is allowed to be an arbitrary commutative ring instead of a field.

Definition 8.8. If R is a ring, the *opposite* of R, denoted R°, is a ring which is equal to R as an additive abelian group, and whose multiplication, $\circ$, is defined by $r \circ s = sr$, where juxtaposition denotes the ordinary multiplication in R. The A-algebra $R \otimes_A R^\circ$ is called the *enveloping algebra* of R. The A-module map $\mu : R \otimes_A R^\circ \to R$ defined on elementary tensors by $\mu(r \otimes s) = rs$ is the *evaluation map*.

Exercise 8.9. Show that defining $(r \otimes s) * t = rts$ makes R into a left $R \otimes_A R^\circ$-module. Show that the evaluation map μ is a homomorphism of left $R \otimes_A R^\circ$-modules. Show that $Ker(\mu)$ is generated as a left ideal in $R \otimes_A R^\circ$ by the set $\{r \otimes 1 - 1 \otimes r \mid r \in R\}$.

Theorem 8.10 ([DI, Prop. 1.1, p. 40]). *Let R be an algebra over the commutative ring A. Make R into a left $R \otimes_A R^\circ$-module by $(r \otimes s) * t = rst$, and let $\mu : R \otimes_A R^\circ \to R$ be the evaluation map. Then the following are equivalent.*

(a) *R is a projective left $R \otimes_A R^\circ$-module.*

(b) *The exact sequence*

$$0 \to Ker(\mu) \longrightarrow R \otimes_A R^\circ \longrightarrow R \to 0$$

   splits as a sequence of $R \otimes_A R^\circ$-modules.

(c) *There exists $e \in R \otimes_A R^\circ$ such that $\mu(e) = 1$ and $Ker(\mu)e = 0$.*

Proof. The equivalence of (a) and (b) is immediate from the definition of projective.

(b) $\Rightarrow$ (c). Let $\varphi : R \to R \otimes_A R^\circ$ be a homomorphism of left $R \otimes_A R^\circ$-modules such that $\mu\varphi$ is the identity map, and let $e = \varphi(1)$. Then $\mu(e) = \mu\varphi(1) = 1$. Moreover, for any $r \in R$, $(r \otimes 1 - 1 \otimes r) * 1 = r - r = 0$. Applying φ shows that $(r \otimes 1 - 1 \otimes r)e = 0$ for all $r \in R$. Hence $Ker(\mu)e = 0$, since $Ker(\mu)$ is generated as a left ideal in $R \otimes_A R^\circ$ by the set $\{r \otimes 1 - 1 \otimes r \mid r \in R\}$.

(c) $\Rightarrow$ (b). If such an e exists, then $\varphi(r) = (r \otimes 1)e$ defines an $R \otimes_A R^\circ$-module homomorphism $\varphi : R \to R \otimes_A R^\circ$ such that $\mu\varphi$ is the identity. $\qquad\square$

Definition 8.11. An A-algebra R over a commutative ring A is a *separable A-algebra* if any of the three equivalent conditions of Theorem 8.10 are satisfied. If A is equal to the center of R, then R is an *Azumaya A-algebra* (or *central separable A-algebra*).

The notion of separability makes sense when A is not the center of R, as, for example, when one considers separability of field extensions. The following result, which will not be used in the sequel, shows that separable extensions can be divided into a central separable part and a separable extension of commutative rings.

Theorem 8.12 ([DI, Theorem 3.8, p. 55]). *Let R be a ring, and suppose that $A \subseteq Z \subseteq R$, where Z is the center of R and A is a subring of Z. Then R is separable over A if and only if R is central separable over Z and Z is separable over A.* $\qquad\square$

Examples 8.13.

1. If A is a field, then R is Azumaya over A if and only if R is a finite-dimensional central simple A-algebra.

2. If A is a commutative ring, $M_n(A)$ is Azumaya over A. More generally, if P is a faithful finitely generated projective A-module, then $End_A(P)$ is Azumaya over A.

3. If R and S are Azumaya algebras over A, so is $R \otimes_A S$.

Exercise 8.14. Show that if A is a field, then R is an Azumaya A-algebra if and only if R is a finite-dimensional central simple A-algebra.

The following surprising result of A. Braun shows that the classical definition of Azumaya algebra can be weakened in a small but significant way. In conjunction wtih central polynomials, it is just what is needed for the quite simple proof of Artin's Theorem in the next section.

Theorem 8.15 (A. Braun [Br1]). *Let R be a ring with center A, let $\mu : R \otimes_A R^\circ \to R$ be the evaluation map, and let R be a left $R \otimes_A R^\circ$-module with action induced by $(r \otimes s) * t = rts$. Then the following are equivalent.*

(a) *R is an Azumaya algebra over A.*

(b) *There is $e \in R \otimes_A R^\circ$ such that $\mu(e) = 1$ and $Ker(\mu)e = 0$.*

(c) *There is $e \in R \otimes_A R^\circ$ such that $\mu(e) = 1$ and $(Ker(\mu)e) * R = 0$.*

(d) *There exist $a_1, \ldots, a_k, b_1, \ldots b_k \in R$ such that $\sum a_i b_i = 1$ and $\sum a_i r b_i \in A$ for all $r \in R$.*

Proof (W. Dicks [Di]). The equivalence of (a) and (b) is the definition of Azumaya algebra, and (b) trivially implies (c).

(c) $\Leftrightarrow$ (d). If $e = \sum a_i \otimes b_i$, then $\mu(e) = \sum a_i b_i$ and $e * r = \sum a_i r b_i$. Noting that $A = \{r \in R \mid Ker(\mu) * r = 0\}$, we have

$$\mu(e) = 1 \text{ and } 0 = (Ker(\mu)e) * R = Ker(\mu) * (e * R)$$
$$\Leftrightarrow \mu(e) = 1 \text{ and } e * R \subseteq A$$
$$\Leftrightarrow \sum a_i b_i = 1 \text{ and } \sum a_i r b_i \in A \text{ for all } r \in R.$$

(c) $\Rightarrow$ (b). Set $U = R \otimes_A R^\circ$, $J = Ker(\mu)$, and suppose that $e \in U$ satisfies $\mu(e) = 1$, $(Je) * R = 0$. Make U into a left U-module by left U-module actions $*_1$ and $*_2$ defined on elementary tensors by $u *_1 (a \otimes b) = (u * a) \otimes b$, $u *_2 (a \otimes b) = a \otimes (u * b)$, where $u \in U$, $a, b \in R$.

We first show that $UeU = U$, where UeU is the two-sided ideal of U generated by e.

Let $N = \{r \in R \mid r \otimes 1 \in UeU\}$, a two-sided ideal of R. Note that $e *_2 (a \otimes b) = a \otimes (e * b) = a(e * b) \otimes 1$ and $U *_2 (UeU) \subseteq UeU$. Thus every element of $e *_2 (UeU)$ is of the form $r \otimes 1$, where $r \in N$.

We claim that $N = R$. Assume not, and let M be a maximal ideal of R which contains N. Consider the composition

$$- : R \otimes_A R^\circ \to (R/M) \otimes_A (R/M)^\circ \to (R/M) \otimes_{\overline{A}} (R/M)^\circ = \overline{U},$$

where $\overline{A}$ is the center of R/M. (The right arrow is an isomorphism because $\overline{A}$ is the image of A in R/M, but we do not need this fact.) Observe that $*_2$ is compatible with $^-$, in the sense that $\overline{u *_2 v} = \overline{u} \,\overline{*_2}\, \overline{v}$ for all $u, v \in U$, where $\overline{*_2}$ is defined analogously to $*_2$. Since R/M is a simple ring, $\overline{U} = (R/M) \otimes_{\overline{A}} (R/M)^\circ$ is a simple ring. Now $0 \neq \overline{1 \otimes 1} = \overline{e *_2 (1 \otimes 1)} = \overline{e} \,\overline{*_2}\, \overline{1 \otimes 1}$. In particular, $\overline{e} \neq 0$, so $\overline{U} = \overline{U}\overline{e}\overline{U} = \overline{UeU}$. Then

$$\overline{e} \,\overline{*_2}\, \overline{U} = \overline{e *_2 U} = \overline{e *_2 (UeU)} \subseteq \overline{N \otimes 1} \subseteq \overline{M \otimes 1} = 0,$$

a contradiction. Thus $N = R$, and $UeU = U$.

We claim that for any $u \in U$, $(UeU) *_2 u \subseteq (u *_1 U)U$. Let $v, w \in U$, let $ew = \sum c_j \otimes d_j$ and let $u = \sum g_k \otimes h_k$. Then

$$
\begin{aligned}
vew *_2 u &= vew *_2 \left(\sum g_k \otimes h_k\right) = \sum g_k \otimes (vew * h_k) \\
&= \sum g_k \otimes (v * 1)(ew * h_k) = \sum g_k(ew * h_k) \otimes (v * 1) \\
&= \sum_{j,k} g_k(c_j h_k d_j) \otimes (v * 1) = \left[\sum_{j,k} g_k c_j h_k \otimes (v * 1)\right](d_j \otimes 1) \\
&= \sum [u *_1 (c_j \otimes (v * 1))](d_j \otimes 1) \subseteq (u *_1 U)U.
\end{aligned}
$$

Since $(Je) * R = 0$ by (c), $(Je) *_1 U = 0$. Hence

$$Je \subseteq U *_2 Je = (UeU) *_2 Je \subseteq ((Je) *_1 U)U = 0,$$

so $Je = 0$, which establishes (b). □

Exercise 8.16. Show that if R is an Azumaya A-algebra and N is a two-sided ideal in R, then the center of R/N is the image of A under the canonical map $R \to R/N$.

We conclude this section by recording some facts about Azumaya algebras which will be useful later.

Theorem 8.17. *Let R be an Azumaya algebra over A. Then*

(a) [DI, Theorem 3.4, p. 52] *The map $R \otimes_A R^\circ \to Hom_A(R, R)$ induced by the *-action of $R \otimes_A R^\circ$ on R, is an isomorphism of A-algebras.*

(b) [DI, Corollary 3.7, p. 54] *Expansion-contraction gives a 1–1 correspondence between ideals of A and two-sided ideals of R. (I.e., For any ideal I of A, $IR \cap A = I$, and for any two-sided ideal J of R, $J = (J \cap A)R$. The 1–1 correspondence preserves sums, products and intersections of ideals, and carries prime ideals to prime ideals.*

(c) [OS, Prop. 2.10, p. 16] *R is a finitely generated projective A-module.*

(d) *For some k there exist orthogonal idempotents $e_1, \ldots, e_k$ of A such that $e_1 + \cdots + e_k = 1$ and each Re_j is Azumaya over constant rank j^2 over Ae_j.*

Proof of (d). Since R is a finitely generated projective A-module, such a decomposition of R into summands Re of constant rank over Ae exists, by Theorem 8.7. This constant rank is the same as the rank of R_P over A_P, where P is any prime ideal of A such that $e \in P$. This in turn is the dimension of R_P/PR_P as a vector space over A_P/PA_P. Since R_P/PR_P is central simple over A/P, this dimension is the square of an integer, and the theorem follows. □

Chapter 9

Artin's Theorem

Artin's Theorem says that if a ring R satisfies the polynomial identities satisfied by $M_n(\mathbb{Z})$ and no nonzero homomorphic image satisfies the polynomial identities satisfied by $M_{n-1}(\mathbb{Z})$, then R is Azumaya. Thus Kaplansky's Theorem, Posner's Theorem and Artin's Theorem have a common theme: If R satisfies a polynomial identity (plus a further hypothesis), then R has a large center (plus a further conclusion).

The original article of M. Artin [Ar] proved the theorem for algebras over a field, and it was extended to rings by C. Procesi [Pr2]. After the discovery of central polynomials, shorter proofs were given by L.H. Rowen [Rw1], S.A. Amitsur [A5], and A.W. Goldie [G]. Their proofs were based on the existence of a central polynomial for $M_n(\mathbb{Z})$ which is linear in one variable and does not vanish on any central simple algebra of dimension n^2 over its center. The shortest proof in a journal article is that of W. Schelter [S1], who used a central polynomial of Capelli type (i.e., a linear combination of Capelli polynomials) to verify an equational definition of Azumaya algebra.

The proof below is approximately as short as Schelter's proof, it is computationally simpler, and it does not require a central polynomial of Capelli type. It uses Braun's characterization of Azumaya algebras (Theorem 8.15) and the following lemma, which occurred (in one form or another) in the papers of Rowen, Amitsur and Goldie cited above.

Lemma 9.1. *Let R be a ring with center A. Suppose that $f(x_1, \ldots, x_k) \in \mathbb{Z}\langle X \rangle$ satisfies*

(1) *f is a central polynomial for R which is linear in x_k.*

(2) *f is not a polynomial identity for any nonzero homomorphic image of R.*

Let $f(R)$ be the abelian group generated by all evaluations of f on R. Then $1 \in f(R)$.

Proof. Assume conversely that $1 \notin f(R)$. By Zorn's Lemma there is an ideal M of R which is maximal among the two-sided ideals of R such that $1 \notin (f(R) + M)$, Note that $1 \in (f(R) + M)$ if and only if $1 \in (f(R) + (M \cap A))$, since $f(R) \subseteq A$.

We claim that M is a prime ideal. If not, there are ideals P, Q of R, strictly larger than M, such that $PQ \subseteq M$. By the maximal choice of M, there are $r, s \in f(R)$, $p \in (P \cap A)$, $q \in (Q \cap A)$, such that $1 = r + p = s + q$. Then $1 = (rs + rq + ps) + pq$, where $(rs + rq + ps) \in f(R)$ and $pq \in M$, a contradiction. Hence M is a prime ideal, as claimed.

Since f is not a polynomial identity for R/M by hypothesis (2), there is an evaluation $e = f(a_1, \ldots, a_{k-1}, d)$ of f which does not lie in M. By the maximality of M, $1 \in (f(R) + (M + Re))$. Letting $^- : R \to R/M$ be the canonical map, this gives an equation

$(*)$ $\qquad\qquad\qquad\qquad\qquad \bar{1} = \bar{a} + \bar{r}\bar{e}$

in R/M, where $a \in f(R)$, $r \in R$. Since R/M is a prime ring, this equation implies that $\bar{r}$ is central in R/M. The linearity of f in the variable x_k implies that $\bar{r}\bar{e} = f(\bar{a}_1, \ldots, \bar{a}_{k-1}, \bar{r}\bar{d}) \in \overline{f(R)}$. Then $(*)$ says that $1 \in f(R) + M$, a contradiction which establishes that $1 \in f(R)$. $\qquad\qquad\qquad\qquad\qquad\qquad\qquad\qquad\Box$

Exercise 9.2. Show that if R is a prime ring with center A, $x, y \in R$, and $x, xy \in A$, then $y \in A$.

Theorem 9.3 (M. Artin [Ar]). *Let R be a ring with center A such that R satisfies all the polynomial identities in $\mathbb{Z}\langle X \rangle$ satisfied by $M_n(\mathbb{Z})$ and no nonzero homomorphic image of R satisfies the polynomial identities satisfied by $M_{n-1}(\mathbb{Z})$. Then R is Azumaya over A, of constant rank n^2.*

Proof. Let $F = F(x, y_1, \ldots, y_n)$ be the central polynomial of Theorem 3.4. By Lemma 4.14(d), it is a central polynomial for any central simple algebra of dimension n^2 over its center.

Suppose that M is a maximal ideal of R. Then R/M is simple, hence primitive, so by Kaplansky's Theorem (4.13) it is central simple of dimension r^2 over its center, for some integer r. By hypothesis, R satisfies the identities satisfied by $M_n(\mathbb{Z})$ and R/M does not satisfy the identities satisfied by $M_{n-1}(\mathbb{Z})$. By Lemma 4.14(b,d), this implies that $r = n$.

Since $F(x, y_1, \ldots, y_n)$ is linear in the variable y_n, it satisfies the hypotheses of Lemma 9.1. Hence there are $r_i, s_{ij} \in R$ ($i = 1, \ldots, m$, $j = 1, \ldots, n$) such that

$$f(r_1, s_{11}, \ldots, s_{1n}) + f(r_2, s_{21}, \ldots, s_{2n}) + \cdots + f(r_m, s_{m1}, \ldots, s_{mn}) = 1.$$

Let $f(x, y_1, \ldots, y_{n-1}, y_n z) = \sum_t a_t(x, y_1, \ldots, y_n) z b_t(x, y_1, \ldots, y_n)$, and set $a_{ti} = a_t(r_i, s_{ij})$, $b_{ti} = b_t(r_i, s_{ij})$. Then $\sum_{t,i} a_{ti} b_{ti} = 1$ and $\sum_{t,i} a_{ti} r b_{ti} \in A$ for all $r \in R$. Thus R is Azumaya over A, by Theorem 8.15.

It is clear that R has constant rank n^2 over A, since for any maximal ideal M of R, R/M has dimension n^2 over its center, which is the image of A in R/M, by exercise 8.16. $\qquad\qquad\qquad\qquad\qquad\qquad\qquad\qquad\qquad\qquad\qquad\Box$

The converse of Theorem 9.3 is also true, and was known much earlier. The following interesting result yields it as an immediate corollary.

Theorem 9.4 ([Rw2, Prop 1.8.46, p. 70]). *Suppose that R is Azumaya over A of constant rank n^2. Then there exists a commutative A-algebra B for which there is an embedding $R \hookrightarrow M_n(B)$.* $\qquad\qquad\qquad\qquad\qquad\qquad\qquad\Box$

Theorem 9.5. *Suppose that R is Azumaya over A of constant rank n^2. Then R satisfies the polynomial identities satisfied by $M_n(\mathbb{Z})$ and no nonzero homomorphic image of R satisfies the polynomial identities satisfied by $M_{n-1}(\mathbb{Z})$.*

Proof. Since $R \hookrightarrow M_n(B)$, where B is commutative, R satisfies all the identities satisfied by $M_n(\mathbb{Z})$. Since R/M is central simple of dimension n^2 over its center whenever M is a maximal ideal of R, no nonzero homomorphic image satisfies the identities satisfied by $M_{n-1}(\mathbb{Z})$, by Lemma 4.14(c).   □

Chapter 10

Chain Conditions

Definition 10.1. A ring R is *left Noetherian* (*right Noetherian, Noetherian*) if it satisfies the ascending chain condition on left ideals (right ideals, two-sided ideals).

The Noetherian condition is fundamental in commutative ring theory. It has been widely studied for noncommutative rings, but generalizing theorems about commutative Noetherian rings to noncommutative Noetherian rings is often difficult, and often is possible only if further conditions are imposed on the ring. The most well-known example of this phenomenon is Goldie's Theorem, which in its basic form gives necessary and sufficient conditions for a prime ring to have a so-called *classical quotient ring*. While Goldie's Theorem is both deep and difficult, when restricted to commutative rings it reduces to a theorem from an undergraduate algebra course: A commutative domain has a quotient field.

Satisfying a polynomial identity is sometimes thought of as a sort of generalized commutativity, and it turns out that a number of results true for commutative rings are also true for PI-rings. The deepest of these are W. Schelter's theorem [S2] that affine PI-algebras over a field are catenary and the Y.P. Razmyslov–A.R. Kemer–A. Braun theorem [R3,Ke,Br2] that the Jacobson radical of an affine PI-algebra over a commutative Noetherian ring is nilpotent. (A ring is *affine* over a commutative ring A if it is finitely generated as an A-algebra.) Both are proved using Noetherian techniques, although affine PI-rings are not Noetherian in general.

In this section we prove several less difficult theorems about chain conditions on PI-rings.

Theorem 10.2. *Let $f(x_1, \ldots, x_k) \in \mathbb{Z}\langle X \rangle$ be a polynomial which is a central polynomial for any central simple algebra of dimension n^2 over its center. (E.g., the polynomial F of Theorem 3.4.) Suppose that R is a prime PI-ring of PI-degree n, and $s = f(r_1, \ldots, r_k)$ is a nonzero evaluation of f on R. Then $R[1/s]$ is Azumaya.*

Proof. Since R has PI-degree n, f is a central polynomial for R. On the other hand, if R/M is a nonzero homomorphic image of R, the image of s is a nonzero evaluation of f on R/M. Hence R/M does not satisfy the polynomial identities of $M_{n-1}(\mathbb{Z})$, since f is a PI for $M_{n-1}(\mathbb{Z})$, by Lemma 3.5. Thus $R[1/s]$ satisfies the hypotheses of Theorem 9.3, so it is Azumaya. $\qquad\square$

The next theorem is a famous result of W. Krull.

Theorem 10.3 (W. Krull [N, p. 60, Theorem 7]). *Let A be a commutative Noetherian ring, and let P be a prime ideal of A. Suppose that P is generated by n elements. Then any properly descending chain of prime ideals $P = P_0 \supset P_1 \supset \cdots \supset P_r$ has length at most n (i.e., $r \leq n$). In particular, A satisfies the descending chain condition on prime ideals.* $\qquad\square$

The next theorem generalizes Theorem 10.3 to PI-rings. It is a model application of Artin's Theorem.

Theorem 10.4 (L.W. Small [S]). *If R is a Noetherian PI-ring, then R satisfies the descending chain condition of prime ideals.*

Proof. Assume conversely that $P_1 \supset P_2 \supset \cdots$ is an infinite properly descending chain of prime ideals in R, where R/P_i has PI-degree n_i. If R satisfies a PI of degree d, then $d \geq 2n_i$ for all i, by Lemma 4.14(c). Hence $n = \max\{n_i\}$ is a finite integer.

Let $P = \bigcap P_i$, and let F_n be the central polynomial for $n \times n$ matrices of Theorem 3.4. It is easy to see that P is a prime ideal and that R/P has PI-degree n. Thus F_n is a central polynomial for the prime PI-ring R/P, so there is an evaluation s of F_n on R such that $s \notin P$. By deleting a finite initial set of P_i, we may suppose that $s \notin P_i$ for all i. Let $S = \{1, s, s^2, \dots\}$, so that $S^{-1}R = R[s^{-1}]$.

Then $S^{-1}P_1 \supset S^{-1}P_2 \supset \cdots$ is a properly descending chain of prime ideals in $S^{-1}R$, which is Azumaya by Theorem 10.2, and Noetherian since R is Noetherian. By Theorem 8.17(b), expansion-contraction gives a 1-1 correspondence between ideals of the center of R and ideals of R, and this correspondence takes prime ideals to prime ideals. The center of R is then a commutative Noetherian ring with an infinite properly descending chain of prime ideals, which contradicts Theorem 10.3. Thus R satisfies the descending chain condition on prime ideals. $\qquad\square$

Recall that if R is an Azumaya algebra with center A, then there is an isomorphism $\epsilon : R \otimes_A R^\circ \to Hom_A(R, R)$ defined on elementary tensors by $\epsilon(a \otimes b)(r) = (a \otimes b) * r = arb$. In other words, if $\theta : R \to R$ is an A-module homomorphism, then there are $a_j, b_j \in R$ such that $\theta(r) = \sum a_j r b_j$ for all $r \in R$.

Theorem 10.5 (E. Formanek [F2]). *Let R be a prime PI-ring of PI-degree n with center A. Then there is an A-module embedding $R \hookrightarrow (A)^{n^2}$.*

Proof. Let $S = A - \{0\}$, and let $S^{-1}R$ be the quotient ring of R given by Posner's Theorem (6.5). Since R has PI-degree n, $S^{-1}R$ has dimension n^2 as a vector space over $S^{-1}A$. Let $r \to (\varphi_1(r), \dots, \varphi_{n^2}(r))$ be an isomorphism between $S^{-1}R$ and $(S^{-1}A)^{n^2}$, where each φ_i is an $(S^{-1}A)$-linear map $S^{-1}R \to S^{-1}A$.

Since $S^{-1}R$ is Azumaya over $S^{-1}A$, the canonical map $S^{-1}R \otimes_{S^{-1}A} (S^{-1}R)^\circ \to Hom_{S^{-1}A}(S^{-1}R, S^{-1}R)$ is an isomorphism (Theorem 8.17(a)). Hence for $i = 1, \dots, n^2$, there are finitely many $a_{ij}, b_{ij} \in S^{-1}R$ such that $\varphi_i(r) = \sum a_{ij} r b_{ij}$ for all $r \in S^{-1}R$. Choose $s \in S$ such that all sa_{ij}, sb_{ij} lie in R (a "common denominator"). Then each $r \to s^2 \varphi_i(r)$ is an A-module map from R to A, and $r \to (s^2 \varphi_1(r), \dots, s^2 \varphi_{n^2}(r))$ is the desired A-module embedding $R \hookrightarrow (A)^{n^2}$. $\qquad\square$

Corollary 10.6 (E. Formanek [F2]). *Let R be a prime PI-ring with center A. If A is Noetherian, then R is a finitely generated A-module, and R is left and right Noetherian.* $\qquad\square$

Theorem 10.7 (Generic flatness for prime PI-rings). *Let R be a prime PI-ring of PI-degree n with center A. Then there is a nonzero $s \in A$ such that $R[1/s]$ is a free module of rank n^2 over $A[1/s]$.*

Proof. Let S and $S^{-1}R$ be as in the proof of Theorem 10.5, and let $r_1, \ldots, r_{n^2} \in R$ be a basis for $S^{-1}R$ over $S^{-1}A$. Then each $r \in S^{-1}R$ is uniquely expressible in the form

$$(*) \qquad\qquad R = \varphi_1(r)r_1 + \cdots + \varphi_{n^2}(r)r_{n^2},$$

where $\varphi_1, \ldots, \varphi_{n^2} \in Hom_{S^{-1}A}(S^{-1}R, S^{-1}A)$. The argument in Theorem 10.5 produces $s \in S$ such that $\varphi_i(R[1/s]) \subseteq A[1/s]$ for all φ_i, and then $(*)$ implies that $R[1/s]$ is free over $A[1/s]$ with basis $r_1, \ldots, r_{n^2}$. $\qquad\square$

The terminology *generic flatness* for the conclusion of Theorem 10.7 is standard, even though *generic freeness* is more logical.

Exercise 10.8. Let E be a field. Show that if R is a subring of $M_n(E)$ which contains an E-basis for $M_n(E)$, then R is a prime PI-ring of PI-degree n.

Example 10.9. The following example, which is taken from [MR, 1.7, p. 12, Example (iii), p. 144, and 6.12, p. 467], shows that the converse of Corollary 10.6 is not true.

Let K be a field, let x, y, t be independent commuting indeterminates, let $K(x, y)$ be the field of quotients of $K[x, y]$, let $R = K(x, y)[t]$, let J be the ideal of R generated by t, and let $A = K(x, y^2) + J$, $B = K(x + y, y^2) + J$. Then A and B are commutative Noetherian rings, and R and J are Noetherian modules over both A and B. The set of 2×2 matrices

$$\begin{pmatrix} A & J \\ J & B \end{pmatrix} = \begin{pmatrix} K(x, y^2) + J & J \\ J & K(x + y, y^2) + J \end{pmatrix}$$

is a prime PI-ring S, by exercise 10.8 applied with $E = K(x, y, t)$, the field of quotients of $K[x, y, t]$. The fact that A and B are Noetherian and J is Noetherian both as an A-module and a B-module implies that S is both left Noetherian and right Noetherian. However, the center of S is $A \cap B = K(y^2) + J$, which is not Noetherian since J is not a finitely generated $K(y^2)$-module.

Theorem 10.10 (G. Cauchon [C]). *Let R be a prime PI-ring. If R is Noetherian (i.e., satisfies the ACC on two-sided ideals), then R is left and right Noetherian.*

Proof. By symmetry, it is enough to prove that R is left Noetherian. It suffices to show that the left ideal of R generated by a sequence of elements $r_1, r_2, \ldots$ is finitely generated. The hypothesis that R satisfies the ACC on two-sided ideals is the same as saying that R satisfies the ACC on R-R-bimodules. A standard argument shows that R^m satisfies the ACC on R-R-bimodules for any positive integer m.

Let $\varphi : R \hookrightarrow R^m$ be the composition of the A-module embedding $R \hookrightarrow A^m$ of Theorem 10.5 with the inclusion $A^m \hookrightarrow R^m$, where $m = n^2$ and n is the PI-degree of R. Since R^m satisfies the ACC on R-R-bimodules, there is an integer k such that the R-R-subbimodule of R^m generated by the $\varphi(r_i)$ is generated by $\varphi(r_i), \ldots, \varphi(r_k)$. This means that for $q > k$, $\varphi(r_q) \in \sum\{R\varphi(r_i)R \mid i = 1, \ldots, k\}$. But since $\varphi(R) \subseteq A^m$, $R\varphi(r_i)R = R\varphi(r_i)$. Since φ is 1-1, this translates to $r_q \in \sum\{Rr_i \mid i = 1, \ldots, k\}$, which is what we needed to prove. $\square$

Exercise 10.11. Show that Corollary 10.6 and Theorem 10.10 hold for semiprime PI-rings.

Chapter 11

Hilbert and Jacobson PI-Rings

Definition 11.1. An algebra over a commutative ring A is *affine over* A if it is finitely generated as an A-algebra. An algebra R over a field K is a *Hilbert algebra over* K if R/M is finite-dimensional over K whenever M is a maximal ideal of R. A ring R is a *Jacobson ring* if every prime ideal is an intersection of maximal ideals.

Notation 11.2. If $y_1, \ldots, y_n$ generate an affine algebra over a commutative ring A, then we write $R = A\{y_1, \ldots, y_n\}$. We use the notation $A[y_1, \ldots, y_n]$ only for a commutative affine algebra (which may or may not be a polynomial ring in the given generators).

The above definitions originate in commutative algebra, and one version of Hilbert's Nullstellensatz is that the commutative polynomial ring $K[y_1, \ldots, y_n]$ over a field K is a Hilbert algebra over K.

In this section we prove theorems of Amitsur and Procesi which generalize theorems about affine commutative rings to affine PI-rings.

Theorem 11.3 (Hilbert's Nullstellensatz).
Let K be a field, and let $L = K[y_1, \ldots, y_n]$ be a commutative affine K-algebra. If L is a field, then L is finite-dimensional over K. $\qquad\square$

We will prove the PI-Nullstellensatz by using the Artin–Tate Lemma to reduce to the commutative Nullstellensatz.

Theorem 11.4 (Artin–Tate Lemma [MR, 9.10, p. 481]). *Let $A \subseteq C \subseteq R$, where A, C, R are rings, C is commutative and R is a C-algebra. (I.e., C is central in R.) Suppose that R is an affine A-algebra and a finitely generated C-module. Then*

(a) *There exists an affine A-subalgebra B of C such that R is a finitely generated B-module.*

(b) *If A is Noetherian, then C is an affine A-algebra.*

Proof. (a) Let $R = A\{r_1, \ldots, r_m\} = Cu_1 + \cdots + Cu_n$. Then there are finitely many $b_{ij}, b_{ijk} \in C$ such that $r_i = \sum_j b_{ij} u_j$, $u_i u_j = \sum_k b_{ijk} u_k$. Then $B = A[b_{ij}, b_{ijk}]$ is an affine A-algebra and $R = Bu_1 + \cdots + Bu_k$.

(b) If A is Noetherian, then B is Noetherian and R is a Noetherian B-module, and so C is a finitely generated B-module. But then C is affine over A, since B is affine over A. $\qquad\square$

Theorem 11.5 (S.A. Amitsur–C. Procesi [AP]). *Let R be an affine PI-algebra over K, where K is a field. Then R is a Hilbert algebra over K.*

Proof. We must show that if R is simple, it is finite-dimensional over K. If R is simple, it is primitive, so by Kaplansky's Theorem (4.13) it is finite-dimensional over its center C. Applying the Artin–Tate Lemma (11.4(b)) to $K \subseteq C \subseteq R$ shows that C is an affine K-algebra. Since C is a field, the commutative Nullstellensatz (11.3) implies that C is finite-dimensional over K. But then R is also finite-dimensional over K, as required.   □

We now use the above theorem to show that under the same hypotheses R is a Jacobson ring.

Exercise 11.6. Suppose that R is a subring of a ring S which is generated over R by central elements. Show that if P is a prime ideal in S, then $P \cap R$ is a prime ideal in R. Give an example which shows that this result fails without the hypothesis that S is generated over R by central elements.

Theorem 11.7 (S.A. Amitsur–C. Procesi [AP]). *Let R be an affine PI-algebra over K, where K is a field. Then R is a Jacobson ring.*

Proof. We must show that any prime ideal P of R is equal to the intersection of the maximal ideals containing it. Passing to R/P, this is equivalent to proving

(∗) If R is a prime affine PI-ring over K, then the intersection of the maximal ideals of R is equal to zero.

To prove (∗), suppose that r is a nonzero element of R, and let t be a central indeterminate over R. By Lemma 6.1, $R[t]$ is semiprimitive, so there is a maximal ideal M of $R[t]$ with $r \notin M$. (Here we implicitly use the fact that primitive ideals of R are maximal ideals, by Kaplansky's Theorem (4.13).) By Theorem 11.5, $R[t]$ is a Hilbert algebra, so $R[t]/M$ is finite-dimensional over K.

By exercise 11.6, $R/(M \cap R)$ is a prime ring. The inclusion of K-algebras $R/(M \cap R) \hookrightarrow R[t]/M$ shows that $R/(M \cap R)$ is finite-dimensional over K. Hence the center of $R/(M \cap R)$ is also finite-dimensional over K, so it is a field. Now $R/(M \cap R)$ is a prime PI-ring whose center is a field, so it is central simple, by Corollary 6.3. Finally, $M \cap R$ is a maximal ideal of R with $r \notin (M \cap R)$, which establishes (∗).   □

Chapter 12

The Ring of Generic Matrices

In Chapter 3, we defined a single $n \times n$ generic matrix. We now introduce the ring generated by a set of $n \times n$ generic matrices.

Definition 12.1. Let K be a field, and let $\{x_{ij}^k \mid 1 \leq i, j \leq n, k \in \mathbb{N}\}$ be independent commuting indeterminates over K. Then

$$X_k = (x_{ij}^k) \in M_n(K[x_{ij}^k])$$

is called an $n \times n$ *generic matrix over* K, and $X_1, X_2, \ldots$ are called *independent generic matrices over* K. The K-subalgebra of $M_n(K[x_{ij}^k])$ generated by $X_1, X_2, \ldots, X_m$ $(m \geq 2)$ or by all the X_k is called a *ring of $n \times n$ generic matrices over* K. The former is denoted $K\{X_1, \ldots, X_m\}$ and the latter is denoted $K\{X\}$.

The reason for the restriction $m \geq 2$ is that when $m = 1$, $K\{X_1\}$ is a polynomial ring in one variable over K, and we want (for $n \geq 2$) the ring of generic matrices to be a noncommutative ring which reflects the noncommutativity of matrix multiplication.

Definition 12.2. Let A be a commutative ring, and let T be a T-ideal in either $A\langle X \rangle$ or in $A\langle x_1, x_2, \ldots, x_m \rangle$. Then $A\langle X \rangle / T$ is called a *relatively free algebra* and $A\langle x_1, x_2, \ldots, x_m \rangle / T$ is called a *relatively free algebra on m generators*.

Theorem 12.3. *Let K be a field, and let $K\{X\}$ be the ring of $n \times n$ generic matrices over K. Then the kernel of the map $K\langle X \rangle \to K\{X\}$ induced by $x_i \to X_i$ is*

$$\mathcal{M}(n) = \{f \in K\langle X \rangle \mid f \text{ is a PI for } M_n(R), \text{ for any commutative } K\text{-algebra } R\}.$$

If K is infinite, then $\mathcal{M}(n)$ is equal to the T-ideal of identities of $M_n(K)$.

Proof. The main assertion of the theorem, that $\mathcal{M}(n)$ is the kernel of $K\langle X \rangle \to K\{X\}$, is an easy exercise which is left to the reader.

For the remaining assertion, note that if K is infinite and L is an extension field of K, then $M_n(K)$ and $M_n(L)$ satisfy the same polynomials in $K\langle X \rangle$, by Theorem 1.9. If R is a commutative K-algebra, then R is a homomorphic image of a polynomial ring over K in sufficiently many variables, and this polynomial ring is a subring of its quotient field. Thus $M_n(K)$ and $M_n(R)$ also satisfy the same polynomials in $K\langle X \rangle$. $\square$

Although Theorem 12.3 is almost a tautology, it is useful to state it explicitly, because it exhibits the relatively free algebra $K\langle X \rangle / \mathcal{M}(n)$ as a concrete object, the ring of $n \times n$ generic matrices over K. Most research about the ring of generic matrices depends on this observation, which Procesi made in 1966.

The important fact that the ring of generic matrices is a domain was proved earlier by Amitsur. His elegant proof depends on Posner's Theorem and the existence of division algebras of dimension n^2 over their centers, which contain K. We precede his proof with a construction of such division algebras.

Theorem 12.4 (S.A. Amitsur [A2]). *Let K be an infinite field, and let J be a nonzero prime T-ideal in $K\langle X \rangle$. Then $J = \mathcal{M}(n)$ for some integer n, where $\mathcal{M}(n)$ is the T-ideal of identities of $M_n(K)$. The T-ideals $\mathcal{M}(n)$ form a properly descending chain $\mathcal{M}(1) \supset \mathcal{M}(2) \supset \mathcal{M}(3) \supset \cdots$.*

Proof. Consider $R = K\langle X \rangle / J$, a prime PI-ring. By Theorem 1.6, $T(R)$, the T-ideal of identities satisfied by R, is equal to J. Since R is a prime PI-ring, Posner's Theorem (6.5) applies, so $T(R) = \mathcal{M}(n)$, where n is the PI-degree of R. Thus $J = T(R) = \mathcal{M}(n)$.

Moreover, since the standard identity s_{2n} is a PI for $M_n(K)$, but s_{2n-2} is not, there are strict inclusions $\mathcal{M}(1) \supset \mathcal{M}(2) \supset \mathcal{M}(3) \supset \cdots$. $\square$

Example 12.5. Let K be a field and let $K[t_1, \ldots, t_n]$ be a commutative polynomial ring over K in n variables, and let $L = K(t_1, \ldots, t_n)$ be its quotient field. Let $\varphi : L \to L$ be the K-algebra automorphism induced by setting $\varphi(t_i) = t_{i+1}$, where indices are taken modulo n. Define a K-algebra $R = (L, \varphi, \sigma^{\pm 1})$ by the following conditions:

(1) As a left L-module, R is free with basis $\{\sigma^i \mid i \in \mathbb{Z}\}$.

(2) Elements of L multiply as usual, and $\sigma^i \sigma^j = \sigma^{i+j}$,

(3) If $a \in L$, $\sigma a = \varphi(a)\sigma$. Thus $(a\sigma^i)(b\sigma^j) = a\varphi^i(b)\sigma^{i+j}$.

Theorem 12.6. *Let $R = (L, \varphi, \sigma^{\pm 1}) = L\{\sigma^{\pm 1}\}$ be defined as above, let C be the center of R, and let $S = C - \{0\}$. Then*

(a) *R has no zero divisors.*

(b) *Let L^φ be the fixed field of L under the action of φ. Then $C = L^\varphi[\sigma^{\pm n}]$, a Laurent polynomial ring over L^φ in one variable.*

(c) *R is a free module over C of rank n^2.*

(d) *$S^{-1}R$ is a division ring of dimension n^2 over its center $S^{-1}C$, and hence R has PI-degree n.*

Proof. (a) Each nonzero element of R can be written in the form $a_i\sigma^i + \cdots + a_j\sigma^j$, where $i, j \in \mathbb{Z}$, $i \leq j$, $a_i, a_j \neq 0$. If we say that such an element has degree j, then the standard degree argument shows that the product of elements of degree j and degree k has degree $j + k$, and hence is nonzero.

(b) By inspection, the centralizer of L in R is $L[\sigma^n]$, and the centralizer of σ in R is $L^\varphi[\sigma]$. The center of R is the intersection of these two centralizers, so

$$C = L[\sigma^n] \cap L^\varphi[\sigma] = L^\varphi[\sigma^n].$$

(c) By Galois theory L has dimension n over L^φ, and by inspection $R = L\{\sigma\}$ is a free module of rank n over $L[\sigma^n]$.

(d) By (c) and Theorem 1.22, R is a PI-ring. Then Posner's Theorem (6.5) applies, so $S^{-1}R$ is central simple over its center, which is $S^{-1}C$. It is a division ring by (a), and it has dimension n^2 over its center, by (c). $\qquad\square$

The division algebra $S^{-1}R$ is a K-algebra which has dimension n^2 over its center. In general, there is no such division algebra whose center is exactly K. For example, if K is algebraically closed, then K itself is the only division algebra with center K which is finite-dimensional over K.

Theorem 12.7 (S.A. Amitsur [A2]). *If K is a field, then the ring of $n \times n$ generic matrices $K\{X\}$ is a domain.*

Proof. Consider $K\{X\} \subseteq M_n(E)$ where $E = K(x_{ij}^k)$ is the quotient field of $K[x_{ij}^k]$. The linear independence of the n^2 matrices $\{X_1^i X_2 X_1^j \mid 1 \le i, j \le n\}$ is equivalent to the nonvanishing of the determinant of the $n^2 \times n^2$ matrix obtained by making a $1 \times n^2$ matrix from the entries of each $n \times n$ matrix $X_1^i X_2 X_1^j$. This determinant does not vanish even if X_1 is specialized to a diagonal matrix (i.e., the off-diagonal entries of X_1 are specialized to zero) by a Vandermonde determinant argument, so it is nonzero. Thus $K\{X\}$ contains an E-basis for $M_n(E)$, so it is a prime PI-ring, by exercise 10.8.

Now suppose conversely that $K\{X\}$ is not a domain. Since it is a prime ring, it contains a nonzero element whose square is zero, by exercise 5.3. The isomorphism $K\{X\} \cong K\langle X\rangle / \mathcal{M}(n)$ implies that there is some $f \in K\langle X\rangle$ such that f is not a PI for $M_n(K)$, but f^2 is.

To complete the proof, let $R = (L, \varphi, \sigma^{\pm 1})$ be the domain constructed in Theorem 12.6. By Posner's Theorem (6.5), $\mathcal{M}(n)$ is the T-ideal of identities satisfied by R. Hence f is a PI for R, but f^2 is not a PI for R. This is a contradiction, since R is a domain. $\qquad\square$

Chapter 13

The Generic Division Ring of Two 2×2 Generic Matrices

Since the ring of generic matrices is a domain (and thus a prime PI-ring) Posner's Theorem (6.5) says that it has a quotient ring which is a division ring.

Definition 13.1. The $n \times n$ *generic division division ring* is the quotient field of the ring of $n \times n$ generic matrices.

The remaining lectures will explore questions about the ring of $n \times n$ generic matrices and the associated generic division ring. Since nothing essential is lost by considering only two generic matrices, we will do this, calling them X and Y.

In general, its structure is only partially understood. However, the generic division ring associated with a pair of 2×2 generic matrices has a particularly simple structure and provides a good introductory example. It was first described by C. Procesi [Pr1], although the underlying calculations had already been made by J.J. Sylvester in 1883 [Sy].

Let K be a field, and let

$$ X = \begin{pmatrix} x_{11} & x_{12} \\ x_{21} & x_{22} \end{pmatrix}, \quad Y = \begin{pmatrix} y_{11} & y_{12} \\ y_{21} & y_{22} \end{pmatrix} $$

be independent generic matrices over K, which means that the entries of X and Y are independent commuting indeterminates over K. Then X and Y are elements of $M_2(K(x_{ij}, y_{ij}))$, and the generic division ring is a K-subalgebra of this matrix ring. Even though the ring of generic matrices is a subring of a matrix ring, it is a domain, by Theorem 12.7.

Consider the following five elements of $K[x_{ij}, y_{ij}]$, where $T(Z)$, $D(Z)$ denote the trace and determinant of a 2×2 matrix Z:

$$ T(X) = x_{11} + x_{22}, \qquad\qquad D(X) = x_{11}x_{22} - x_{12}x_{21}, $$
$$ T(Y) = y_{11} + y_{22}, \qquad\qquad D(Y) = y_{11}y_{22} - y_{12}y_{21}, $$
$$ T(XY) = x_{11}y_{11} + x_{12}y_{21} + x_{21}y_{12} + x_{22}y_{22}. $$

It can be verified that the above five elements are algebraically independent over K, so $F = K(T(X), T(Y), D(X), D(Y), T(XY))$ is a *rational function field over* K (or a *pure transcendental extension of* K) of transcendence degree five.

Now $F \subset K(x_{ij}, y_{ij})$ and, as noted above, $X, Y \in M_n(K(x_{ij}, y_{ij}))$. When appropriate, we identity $a \in F$ with the scalar matrix $a \cdot I$. The nonvanishing of a certain 4×4 determinant shows that I, X, Y, XY are linearly independent over $K(x_{ij}, y_{ij})$, and hence over F. Now let

$$D = F \cdot I \oplus F \cdot X \oplus F \cdot Y \oplus F \cdot XY$$

be the indicated four-dimensional vector space over F. Calculations based on the two identities

$$Z^2 = T(Z) \cdot Z - D(Z) \cdot I \text{ (for any } 2 \times 2 \text{ matrix } Z),$$
$$YX = (T(XY) - T(X)T(Y)) \cdot I + T(Y) \cdot X + T(X) \cdot Y - XY$$

show that the product of any pair of I, X, Y, XY lie in D. Thus D is a ring which contains X and Y. The center of D is $F \cdot I$, which is a field, so D is central simple. It is a concrete model of the generic division ring. We summarize this as

Theorem 13.2 (C. Procesi [Pr1]). *Let $K\{X, Y\}$ be the ring of two 2×2 generic matrices over K, and let $K(X, Y)$ be its quotient division ring. Then the center of $K(X, Y)$ is the above field F, a rational function field over K in five variables. Moreover I, X, Y, XY are a basis for $K(X, Y)$ over F.* ☐

The above example is helpful in providing a picture of the generic division ring, but it is misleading in that it has many pleasant properties which are absent for larger n. For example, it is true for any n that the matrices $\{X^i Y^j, 0 \leq i, j \leq n - 1\}$ form a basis for the generic division ring over its center, but only for $n = 2$ does the formula which expresses YX as a linear combination of these basis elements have an expression "without denominators" as it does for $n = 2$.

In the sequel, we will explore the question of whether the center of the generic division ring is a rational function field over K. No counterexamples are known, but only a few positive results have been obtained. Moreover, there is no case for $n > 2$ where a transcendence base for the center over K has such a simple form as it does for $n = 2$.

Chapter 14

The Center of the Generic Division Ring

The goal of this section is to give a description of the center of the generic division ring as the fixed field of the symmetric group acting on a rational function field.

Let $X = (x_{ij})$, $Y = (y_{ij})$ be $n \times n$ generic matrices, where the entries are independent commuting indeterminates over the field K. Let E be an algebraic closure of $K(x_{ij})$. Since X has distinct characteristic values, it is diagonalizable over E, so there is a matrix $Z \in M_n(E)$ such that ZXZ^{-1} is diagonal. Set

$$
U = ZXZ^{-1} = \begin{pmatrix} u_1 & & & \\ & u_2 & & \\ & & \ddots & \\ & & & u_n \end{pmatrix},
$$

$$
V = ZYZ^{-1} = \begin{pmatrix} v_{11} & v_{12} & \cdots & v_{1n} \\ v_{21} & v_{22} & \cdots & v_{2n} \\ \vdots & \vdots & & \vdots \\ v_{n1} & v_{n2} & \cdots & v_{nn} \end{pmatrix}.
$$

Lemma 14.1. *Conjugation by the above Z defines a K-algebra isomorphism between the ring of $n \times n$ generic matrices $K\{X, Y\}$ and $K\{U, V\}$, the K-subalgebra of $M_n(K(u_i, v_{ij}))$ generated by the above U and V. The nonzero entries of U and V are algebraically independent over K.* $\qquad\square$

Set $L = K(u_i, v_{ij})$, and let $K(U, V)$ be the ring of quotients of $K\{U, V\}$ obtained via Posner's Theorem. Thus $K\{U, V\}$ is (K-isomorphic to) the $n \times n$ generic division ring. The description of the center of the generic division ring which follows is due to C. Procesi [Pr1] as refined by E. Formanek [F3].

We regard $K(U, V)$ as the K-subalgebra of $M_n(L)$ generated by U and V. The center of $K(U, V)$ is then a subfield of L, where $a \in L$ is identified with the scalar matrix $a \cdot I$. Let

$$
B = \langle u_1, \ldots, u_n, v_{11}, v_{12}, \ldots, v_{nn} \rangle, \quad P = \langle p_1, \ldots, p_n \rangle, \quad Q = \langle q \rangle
$$

be multiplicative free abelian groups on the given generators. Then B is a subgroup of the multiplicative group of L, and generates L as a field over K. In other words,

L is the quotient field of the group ring (or Laurent polynomial ring) $K[B] = K[u_i^{\pm 1}, v_{ij}^{\pm 1}]$. The symmetric group S_n acts on B and P by permuting indices, and we let S_n act trivially on Q and K. Since B generates L as a field over K, this gives an action of S_n as K-automorphisms of L. The next lemma is a series of straightforward computations.

Lemma 14.2 ([F3, p. 206]). *Define S_n-homomorphisms $\alpha : B \to P$ and $\beta : P \to Q$ by $\alpha(u_i) = 1$, $\alpha(v_{ij}) = p_i p_j^{-1}$, $\beta(p_i) = q$, and let $A = ker(\alpha)$. Then*

$$1 \to A \longrightarrow B \longrightarrow P \longrightarrow Q \to 1$$

is an exact sequence of S_n-modules. The group A is free abelian of rank $n^2 + 1$ and the subfield $K(A)$ of L generated by A over K is a rational function field over K of transcendence degree $n^2 + 1$. The abelian group A is generated by $u_1, \ldots, u_n$ and all monomials in the v_{ij} of the form $v_{i_1 i_2} v_{i_2 i_3} \cdots v_{i_r i_1}$. The action of S_n on A induces an action of S_n as a group of K-automorphisms of $K(A)$. □

Theorem 14.3 (C. Procesi [Pr1]; [F3]; [He2, p. 20]). *The center C of the generic division ring $K(U, V)$ is the fixed field of S_n acting as indicated above on $K(A)$, a rational function field over K of transcendence degree $n^2 + 1$.*

Proof. We first claim that $C \subseteq K(A)^{S_n}$. Since C is the quotient field of the center of $K\{U, V\}$ by Posner's Theorem (6.5), it is enough to show that $K(A)^{S_n}$ contains the center of $K\{U, V\}$. This is a consequence of the way the matrices U and V multiply. For if $w = (w_{ij}) \in K\{U, V\}$, then the (ij)-th entry of w is a K-linear combination of monomials of the form

$$u_1^{a_1} \cdots u_n^{a_n} v_{ii_2} v_{i_2 i_3} \cdots v_{i_r j}.$$

Hence the diagonal entries of w lie in $K(A)$. Moreover, S_n acts on the entries of w by $\pi(w_{ij}) = w_{\pi(i)\pi(j)}$, and so permutes transitively the diagonal entries of w. Thus, if $w = w_0 \cdot I$ lies in the center of $K\{U, V\}$, then $w_0 \in K(A)$ and S_n fixes w_0. In other words, the center of $K\{U, V\}$ – and hence C – is contained in $K(A)^{S_n}$.

In order to obtain the equality $K(A)^{S_n} = C$, consider the algebra $D = K(U, V) \otimes_C C(u_1, \ldots, u_n)$ obtained by adjoining the characteristic roots of U to C. As a subring of $M_n(L)$, D is generated by $K(U, V)$ and the scalar matrices $u_1 \cdot I, \ldots, u_n \cdot I$. The center of D is $C(u_1, \ldots, u_n)$, and D is central simple of dimension n^2 over $C(u_1, \ldots, u_n)$. Since D contains the n orthogonal idempotents

$$e_{jj} = \frac{\prod_{i \neq j}(U - u_i \cdot I)}{\prod_{i \neq j}(u_j - u_i)}, \quad (j = 1, \ldots, n),$$

D is isomorphic to $M_n(C(u_1, \ldots, u_n))$.

We claim that $C(u_1, \ldots, u_n) \supseteq K(A)$. By Lemma 14.3, A is generated as a field over K by $u_1, \ldots, u_n$ and all monomials in the v_{ij} of the form

(*) $v_{i_1 i_2} v_{i_2 i_3} \cdots v_{i_r i_1}.$

to see that $C(u_1, \ldots, u_n)$ contains all monomials (*), note that

$$v_{ij}e_{ij} = e_{ii}Ve_{jj} \in D,$$
$$(v_{i_1i_2}v_{i_2i_3}\cdots v_{i_ri_1})e_{i_1i_1} = (v_{i_1i_2}e_{i_1i_2})(v_{i_2i_3}e_{i_2i_3})\cdots(v_{i_ri_1}e_{i_ri_1}) \in D,$$
$$(v_{i_ji_{j+1}}\cdots v_{i_{j-1}i_j})e_{i_ji_j} \in D.$$

Thus $(v_{i_1i_2}v_{i_2i_3}\cdots v_{i_ri_1})\cdot I \in D$, provided that $1, \ldots, n$ all occur among the indices $i_1, \ldots, i_r$. Since an arbitrary monomial of the form (*) is a quotient of two such monomials, D contains all monomials (*). Hence $K(A) \subseteq C(u_1, \ldots, u_n)$.

Finally, consider the inclusions

$$C \subseteq K(A)^{S_n} \subseteq K(A) \subseteq C(u_1, \ldots, u_n).$$

Since $u_1, \ldots, u_n$ are the roots of a polynomial of degree n with coefficients in C, $[C(u_1, \ldots, u_n) : C] \leq n!$. Thus

$$
\begin{aligned}
n! \geq\ & [C(u_1, \ldots, u_n) : C] \\
=\ & [C(u_1, \ldots, u_n) : K(A)]\,[K(A) : K(A)^{S_n}]\,[K(A)^{S_n} : C] \\
=\ & [C(u_1, \ldots, u_n) : K(A)]\,n!\,[K(A)^{S_n} : C].
\end{aligned}
$$

This forces $C = K(A)^{S_n}$, as required. □

The next lemma shows why the $n \times n$ generic division ring generated by $r \geq 2$ generic matrices is not essentially different from that generated by two generic matrices.

Lemma 14.4 (C. Procesi [Pr1]). *If $r \geq 3$, the center of the $n \times n$ generic division ring $K(X_1, \ldots, X_r)$ generated by r independent generic matrices $X_1, \ldots, X_r$ is a rational function field in $(r-2)n^2$ variables over the center of $K(X_1, X_2)$. If $Z_1, \ldots, Z_{n^2}$ is any basis for $K(X_1, X_2)$ over its center, then the $(r-2)n^2$ traces $T(Z_iX_j)$ $(i = 1, \ldots, n^2, j = 3, \ldots, r)$ may be taken as the $(r-2)n^2$ variables.* □

Definition 14.5. An extension field of a field K is *unirational* over K if it is contained (as a K-algebra) in a rational function field over K.

Combining Theorems 14.3 and 14.4 gives

Theorem 14.6 (C. Procesi [Pr1]; A.A. Kirillov [Ki]; [He2, p. 19]). *The center of the $n \times n$ generic division algebra $K(X_1, \ldots, X_r)$ generated by $r \geq 2$ $n \times n$ generic matrices is unirational over K of transcendence degree $(r-1)n^2 + 1$.* □

Chapter 15

Is the Center of the Generic Division Ring a Rational Function Field?

Theorem 14.6 says that the center of the generic division algebra $K(U,V)$ is unirational over the base field K, but is it actually a rational function field over K? Here we survey the small number of positive results and discuss the significance of this question. We use the description of the center given by Theorem 14.3.

Example 15.1. $n = 2$. Here the center is $K(A)^{S_2}$, where

$$K(A) = K(u_1, u_2, v_{11}, v_{22}, v_{12}v_{21}).$$

We claim that

$$
\begin{aligned}
K(A)^{S_2} &= K(u_1 + u_2, u_1u_2, v_{11} + v_{22}, v_{11}v_{22} - v_{12}v_{21}, u_1v_{11} + u_2v_{22}) \\
&= K(T(U), D(U), T(V), D(V), T(UV)) = F,
\end{aligned}
$$

(where $T(Z), D(Z)$ are the trace and determinant of Z), which we know is the correct answer, by Theorem 13.2. To verify that the indicated field F is the fixed field of S_2, we first observe that F is fixed by S_2, and then that the quadratic extension $F(u_1)$ of F is equal to $K(A)$. Then the inclusions

$$F \subseteq K(A)^{S_2} \subseteq K(A) = F(u_1),$$

combined with $[K(A) : K(A)^{S_2}] = 2 = [F(u_1) : F]$ implies that $F = K(A)^{S_2}$.

The case $n = 3$ is more difficult, but still can be done by hand. Note, however, that fractions now appear.

Theorem 15.2 (E. Formanek [F3]). *The center of the 3×3 generic division algebra $K(U,V)$ is a rational function field in 10 variables.*

Proof. The abelian group A of Lemma 14.2 is free of rank 10, and

$$A = \langle u_1, u_2, u_2, v_{11}, v_{22}, v_{33}, v_{12}v_{21}, v_{23}v_{32}, v_{31}v_{13}, v_{12}v_{23}v_{31} \rangle$$

exhibits a free generating set. Set

$$a_1 = u_1 + u_2 + u_3, \quad a_2 = u_1u_2 + u_2u_3 + u_3u_1, \quad a_3 = u_1u_2u_3,$$

$$a_4 = v_{11} + v_{22} + v_{33}, \quad a_5 = u_1v_{11} + u_2v_{22} + u_3v_{33}, \quad a_6 = u_1^2v_{11} + u_2^2v_{22} + u_3^2v_{33},$$

$$b = v_{12}v_{21} + v_{23}v_{32} + v_{31}v_{13}, \quad w_1 = v_{23}v_{32}b^{-1}, \quad w_2 = v_{31}v_{13}b^{-1}, \quad w_3 = v_{12}v_{21}b^{-1},$$

$$a_7 = u_1w_1 + u_2w_2 + u_3w_3, \quad a_8 = u_1^2w_1 + u_2^2v_2 + u_3^2w_3,$$

$$p = v_{12}v_{23}v_{31}b^{-1}, \quad q = v_{13}v_{32}v_{21}b^{-1},$$

$$a_9 = p + q, \quad a_{10} = (u_1^2u_2 + u_2^2u_3 + u_3^2u_1)p + (u_1u_2^2 + u_2u_3^2 + u_3u_1^2)q.$$

We will show that $K(a_1, \ldots, a_{10}) = K(A)^{S_3}$. Set $F = K(a_1, \ldots, a_{10})$. By inspection, $F \subseteq K(A)^{S_3}$.

Clearly $F(u_1, u_2, u_3) \subseteq K(A)$. We claim that they are equal. Note that the determinants

$$\det \begin{pmatrix} 1 & 1 & 1 \\ u_1 & u_2 & u_3 \\ u_1^2 & u_2^2 & u_3^2 \end{pmatrix}, \quad \det \begin{pmatrix} 1 & 1 \\ u_1^2 u_2 + u_2^2 u_3 + u_3^2 u_1 & u_1 u_2^2 + u_2 u_3^2 + u_3 u_1^2 \end{pmatrix}$$

are nonzero. Thus $v_{11}, v_{22}, v_{33} \in F(u_1, u_2, u_3)$, since $a_4, a_5, a_6 \in F$. Similarly, $w_1, w_2, w_3 \in F(u_1, u_2, u_3)$, since $1 = w_1 + w_2 + w_3, a_7, a_8 \in F$, and $p, q \in F(u_1, u_2, u_3)$, since $a_9, a_{10} \in F$. But then

$$b = pq(w_1 w_2 w_3)^{-1} \in F(u_1, u_2, u_3)$$

as well, after which it is easily seen that all 10 of the given generators of A are in $F(u_1, u_2, u_3)$. Thus $F(u_1, u_2, u_3) = K(A)$, as claimed.

Finally, $[K(A) : F] = [F(u_1, u_2, u_3) : F] \leq 6$, since F contains the elementary symmetric functions in u_1, u_2, u_3, and $[K(A) : K(A)^{S_3}] = 6$, by Galois theory. Then the inclusions

$$F \subseteq K(A)^{S_3} \subset K(A) = F(u_1, u_2, u_3)$$

show that $K(A)^{S_3} = F = K(a_1, \ldots, a_{10})$, completing the proof. □

In order to discuss further results on the center of the ring of generic matrices, we need a definition.

Definition 15.3. A field F containing a field K is *stably rational* over K if for some r the field $F(u_1, \ldots, u_r)$ is a rational function field over K, where $u_1, \ldots, u_r$ are independent variables over L.

Two fields E and F are *stably birational* if there are independent variables $u_1, \ldots, u_r, v_1, \ldots, v_s$ such that $E(u_1, \ldots, u_r)$ and $F(v_1, \ldots, v_s)$ are isomorphic.

There exist stably rational fields which are not rational The first examples over an algebraically closed field K were given in 1985 [BCSS].

In my opinion, the question of the stable rationality of the center of the generic division ring is the most important open question in PI-theory. A positive solution would imply the Merkurjev–Suslin Theorem [MS; Rw3, p. 198] that the Brauer group of a field containing all roots of unity is generated by cyclic algebras. A. Schofield [Sc2] has shown that the center of the $n \times n$ generic division ring is stably birational to the function fields associated with many moduli spaces. Thus a positive or negative solution would answer important open questions in algebraic geometry.

So far, there are no counterexamples to the rationality of the center, but the positive results are few: The center is rational for $n = 1$ (trivial), $n = 2$ (C. Procesi [Pr1]), $n = 3, 4$ (E. Formanek [F3,F4]), and stably rational for $n = 5, 7$ (C. Bessenrodt–L. LeBruyn [BL]). In addition, P.I. Katsylo [Ka] and A. Schofield [Sc1] have shown that there is stable rationality for $n = rs$ if there is stable rationality for $n = r$ and $n = s$, with r and s relatively prime. Thus the center of the $n \times n$ generic division algebra over a field K is stably rational over K if n is a divisor of $420 = 4 \cdot 3 \cdot 5 \cdot 7$.

The positive results have come with increasing complexity. For $n = 2$, the fixed field problem could be solved "without denominators" but this is no longer possible for $n \geq 3$. For $n \leq 3$, the coefficients of the characteristic polynomial of one of the generic matrices could be taken to be part of generating set, but this is no longer possible for $n \geq 4$. For $n = 4, 5, 7$, the problem was solved by reducing it to the question of the stable rationality of a field of transcendence degree $n - 1$ over K, rather than the field of transcendence degree $n^2 + 1$ given by Theorem 14.3, but this is no longer possible for larger prime n.

The only new idea to appear since 1980 is E. Beneish's attempt to replace, for n prime, the action of S_n by the action of the normalizer of a Sylow n-subgroup of S_n. This has not established stable rationality in any new cases, but it has led to simplified proofs of stable rationality for $n = 5$ and $n = 7$ [B].

Bibliography

[A1]   S.A. Amitsur, *The identities of PI-rings*, Proc. Amer. Math. Soc. **3** (1952), 27–34.

[A2]   S.A. Amitsur, *The T-ideals of the free ring*, J. London Math Soc. **30** (1955), 470–475.

[A3]   S.A. Amitsur, *Radicals of polynomial rings*, Canadian J. Math. **8** (1956), 355–361.

[A4]   S.A. Amitsur, *Rational identities and applications to algebra and geometry*, J. Algebra **3** (1966), 304–359.

[A5]   S.A. Amitsur, *Polynomial identities and Azumaya algebras*, J. Algebra **27** (1973), 117–125.

[AL]   S.A. Amitsur and J. Levitzki, *Minimal identities for algebras*, Proc. Amer. Math. Soc. **2** (1950), 449–463.

[AP]   S.A. Amitsur and C. Procesi, *Jacobson-rings and Hilbert algebras with polynomial identities*, Annali di Mathematica Pura et Applicata **71** (1966), 61–72.

[Ar]   M. Artin, *On Azumaya algebras and finite-dimensional representations of rings*, J. Algebra **11** (1969), 523–563.

[AM]   M.F. Atiyah and I.G. Macdonald, *Introduction to Commutative Algebra*, Addison-Wesley, 1969.

[BCSS]   A. Beauville, J.-L. Colliot-Thélène, J.-J. Sansuc and P. Swinnerton-Dyer, *Variétés stablement rationnelles non rationnelles*, Ann. of Math. **121** (1985), 283–318.

[B]   E. Beneish, *Induction theorems on the stable rationality of the ring of generic matrices*, Trans. Amer. Math. Soc. **350** (1998), 3571–3585.

[BL]   C. Bessenrodt and L. LeBruyn, *Stable rationality of certain PGL_n quotients*, Invent. Math. **104** (1991), 179–199.

[Br1]   A. Braun, *On Artin's Theorem and Azumaya algebras*, J. Algebra **77** (1982), 323–332.

[Br2]   A. Braun, *The nilpotency of the radical in a finitely generated PI-ring*, J. Algebra **89** (1984), 375–396.

[C]   G. Cauchon, *Anneaux semi-premiers, Noethériens, à identités polynomiales*, Bull. Soc. Math. France **104** (1976), 99–111.

[D]   M. Dehn, *Über die Grundlagen der projektiven Geometrie und allgemeine Zahlsysteme*, Math. Ann. **85** (1922), 184–193.

[DI]   F. DeMeyer and E. Ingraham, *Separable Algebras over Commutative Rings*, Springer-Verlag, 1971.

[Di]   W. Dicks, *On a characterization of Azumaya algebras*, Publicacions Mat. **32** (1988), 165–166.

[DK]   V. Drensky and A. Kasparian, *A new central polynomial for* 3×3 *matrices*, Comm. Algebra **13** (1985), 745–752.

[DP]   V. Drensky and G.M. Piacentini-Cattaneo, *A central polynomial of low degree for* 4×4 *matrices*, J. Algebra **168** (1994), 469–478.

[F1]   E. Formanek *Central polynomials for matrix rings*, J. Algebra **23** (1972), 129–132.

[F2]   E. Formanek, *Noetherian PI-rings*, Comm. Alg. **1** (1974), 79–86.

[F3]   E. Formanek, *The center of the ring of* 3×3 *generic matrices*, Linear and Multilinear Algebra **7** (1979), 203–212.

[F4]   E. Formanek, *The center of the ring of* 4×4 *generic matrices*, J. Algebra **62** (1980), 304–319.

[G]   A. Goldie, *Azumaya algebras and rings with polynomial identity*, Math. Proc. Cambridge Phil. Soc. **79** (1976), 393–399.

[H]   M. Hall, *Projective planes*, Trans. Amer. Math. Soc. **54** (1943), 229–277.

[He1]   I.N. Herstein, *Noncommutative Rings*, Math. Association of America, 1968.

[He2]   I.N. Herstein, *Notes from a Ring Theory Conference*, CBMS Regional Conf. Series in Math. No. 78, Amer. Math. Soc., 1970.

[J]   N. Jacobson, *Basic Algebra II*, Second Edition, W. H. Freeman and Co., 1989.

[K]   I. Kaplansky, *Rings with a polynomial identity*, Bull. Amer. Math. Soc. **54** (1948), 575–580.

[Ka]   P.I Katsylo, *Stable rationality of fields of invariants of linear representations of the groups* PSL_6 *and* PSL_{12}, Math. Zametki **48** (1990), 49–52 (Russian). Translation: Math. Notes **48** (1991), 751–753.

[Ke]   A.R. Kemer, *Capelli identities and nilpotency of the radical in finitely generated PI-algebras*, Dokl. Akad. Nauk. SSSR **255** (1980), 793–797 (Russian). Translation: Soviet Math. Dokl. **22** (1980).

[Ki]   A.A. Kirillov, *Certain division algebras over a field of rational functions*, Funktsional Analiz i Prilozhen **1** (1967), 101–102 (Russian).

[Ko] B. Kostant, *A theorem of Frobenius, a theorem of Amitsur-Levitski, and cohomology theory*, J. Math. Mech. **7** (1958), 237–264.

[L] J. Levitzki, *A theorem on polynomial identities*, Proc. Amer. Math. Soc. **1** (1950), 334–341.

[M] I.G. Macdonald, *Symmetric Functions and Hall Polynomials*, Second Edition, Oxford, 1995.

[MR] J.C. McConnell and J.C. Robson, *Noncommutative Noetherian Rings*, Wiley-Interscience, 1987.

[MS] A.S. Merkurjev and A.A. Suslin, *K-cohomology of Severi-Brauer varieties and the norm residue homomorphism*, Izv. Akad. Nauk. SSSR **46** (1982), 1011–1046 (Russian). Translation: Math. USSR Izv. **21** (1983), 307–340.

[N] D.G. Northcott, *Ideal Theory*, Cambridge University Press, 1968.

[OS] M. Orzech and C. Small, *The Brauer Group of Commutative Rings*, Marcel Dekker, 1975.

[P] D.S. Passman, *The Algebraic Structure of Group Rings*, Wiley-Interscience, 1977.

[Po] E.C. Posner, *Prime rings satisfying a polynomial identity*, Proc. Amer. Math. Soc. **11** (1960), 180–184.

[Pr1] C. Procesi, *Non-commutative affine rings*, Atti. Accad. Naz. Lincei Rend. Cl. Sci. Fis. Mat. Natur. **8** (1967), 239–255.

[Pr2] C. Procesi, *On a theorem of M. Artin*, J. Algebra **22** (1972), 309–315.

[Pr3] C. Procesi, *The invariant theory of $n \times n$ matrices*, Adv. in Math. **19** (1976), 306–381.

[R1] Y.P. Razmyslov, *On a problem of Kaplansky*, Izv. Akad. Nauk SSSR **37** (1973), 483–501 (Russian). Translation: Math. USSR-Izv. **7** (1973), 479–496.

[R2] Y.P. Razmyslov, *Trace identities of full matrix algebras over a field of characteristic zero*, Izv. Akad. Nauk SSSR **38** (1974), 723–756 (Russian). Translation: Math. USSR-Izv. **8** (1974), 727–760.

[R3] Y.P. Razmyslov, *The Jacobson radical in PI-algebras.* Alg. i Logika **13** (1974), 192–204 (Russian). Translation: Algebra and Logic **13** (1974), 291–204.

[Re] A. Regev, *the representations of S_n and explicit identities for P.I. Algebras*, J. Algebra **51** (1978), 25–40.

[Ro] S. Rosset, *A new proof of the Amitsur-Levitski identity*, Israel J. Math. **23** (1976), 187–188.

[Rw1]   L.H. Rowen, *On rings with central polynomials*, J. Algebra **31** (1974), 393–426.

[Rw2]   L.H. Rowen, *Polynomial Identities in Ring Theory*, Academic Press, 1980.

[S1]   W. Schelter, *Azumaya algebras and Artin's theorem*, J. Algebra **46** (1977), 303–304.

[S2]   W. Schelter, *Noncommutative affine PI rings are catenary*, J. Algebra **51** (1978), 12–18.

[Sc1]   A. Schofield, *Matrix invariants of composite size*, J. Algebra **147** (1992), 345–349.

[Sc2]   A. Schofield, *Birational classification of moduli spaces*, pp. 297–309 in *Infinite Length Modules*, H. Krause and C.M. Ringel, Eds. Birkhäuser, 2000.

[Sm]   L.W. Small, *Prime ideals in Noetherian PI-rings*, Bull. Amer. Math. Soc. **79** (1973), 421–422.

[Sy]   J.J. Sylvester. *On the involution of two matrices of the second order*, British Association Report, Southport, 1883, pp. 430-432. Pp. 115–117 in: *Collected Mathematical Papers*, Vol 4, Chelsea, 1973.

[W]   W. Wagner, *Über die Grundlagen der projektiven Geometrie und allgemeine Zahlsysteme*, Math. Z. **113** (1937), 528–567.

Index

affine, 173

affine algebra, 177

algebra of invariants, 59

alternating, 141

Amitsur, S.A., 140, 143, 151, 155, 158, 159, 161, 169, 178, 180, 181

Amitsur–Levitzki theorem, 37, 133, 143, 145

antichain, 103

Artin's theorem, 133, 165, 169, 170, 174

Artin, M., 169, 170

Artin–Tate lemma, 177, 178

associated trace function, 63

Azumaya algebra, 133, 163–165, 170, 173, 174

basis of identities of algebra, 8

basis of T-ideal, 8

Beneish, E., 191

Bergman gap theorem, 99

Bessenrodt, C., 191

bimodule, 152

Birkhoff theorem, 115

bounded degree, 141

branching theorem, 28

Brauer group, 190

Braun's characterization of Azumaya algebras, 165

Braun, A., 133, 165, 173

Capelli identity, 6

Capelli polynomial, 141, 169

Cauchon, G., 175

Cayley–Hamilton theorem, 143

central polynomial, 49, 133, 143, 147, 148, 160, 165, 169, 170, 173, 174

central separable, 165

central simple algebra, 153, 160, 170, 173

chain, 103

Chevalley–Jacobson density theorem, 151, 152

cocharacter, 32

codimension, 19

codimension sequence, 19

codimension series, 24

commutator, 6

    long commutator, 6

comparable elements, 103

complexity, 90

complexity function, 24

consequence of identities, 8

constant rank, 164, 170

decomposable word, 87

Dehn, M., 151

dense ring of linear transformations, 152, 153

density theorem, 151, 152

derivation, 84

diagram, *see* Young diagram

Dicks, W., 166

Dilworth theorem, 103

double Capelli identity, 40

double staircase, 40

Drensky, V.S., 143

elementary symmetric function, 144

enveloping algebra, 164

equivalent polynomial identities, 8

Eulerian graph, 37

evaluation map, 164

exponential codimension series, 24

exponential generating function, 23

exterior algebra, 7, 140, 143, 147

extremal variety, *see* minimal variety